Marine Molecular Biotechnology

Subseries of Progress in Molecular and Subcellular Biology
Series Editor: Werner E. G. Müller

Progress in Molecular and Subcellular Biology **42**
Series Editors: W.E.G. Müller (Managing Editor)
Ph. Jeanteur, Y. Kuchino, A. Macieira-Coelho, R. E. Rhoads

Volumes Published in the Series

Progress in Molecular and Subcellular Biology

Subseries: Marine Molecular Biotechnology

Volume 27
Signaling Pathways for Translation: Stress, Calcium, and Rapamycin
R.E. Rhoads (Ed.)

Volume 28
Small Stress Proteins
A.-P. Arrigo and W.E.G. Müller (Eds.)

Volume 29
Protein Degradation in Health and Disease
M. Reboud-Ravaux (Ed.)

Volume 30
Biology of Aging
A. Macieira-Coelho

Volume 31
Regulation of Alternative Splicing
Ph. Jeanteur (Ed.)

Volume 32
Guidance Cues in the Developing Brain
I. Kostovic (Ed.)

Volume 33
Silicon Biomineralization
W.E.G. Müller (Ed.)

Volume 34
Invertebrate Cytokines and the Phylogeny of Immunity
A. Beschin and W.E.G. Müller (Eds.)

Volume 35
RNA Trafficking and Nuclear Structure
Dynamics
Ph. Jeanteur (Ed.)

Volume 36
Viruses and Apoptosis
C. Alonso (Ed.)

Volume 38
Epigenetics and Chromatin
Ph. Jeanteur (Ed.)

Vol. 40
Developmental Biology of Neoplastic Growth
A. Macieira-Coelho (Ed.)

Vol. 41
Molecular Basis of Symbiosis
J. Overmann (Ed.)

Volume 37
Sponges (Porifera)
W.E.G. Müller (Ed.)

Volume 39
Echinodermata
V. Matranga (Ed.)

Volume 42
Antifouling Compounds
N. Fusetani and A.S. Clare (Eds.)

Nobuhiro Fusetani   Anthony S. Clare (Eds.)

Antifouling Compounds

With 45 Figures, 14 in Color, and 15 Tables

Springer

Professor Dr. Nobuhiro Fusetani
Faculty of Fisheries Sciences
Hokkaido University
3-1-1 Minato-cho
Hakodate 041-8611
Japan

E-Mail: anobu@fish.hokudai.ac.jp

Professor Dr. Anthony S. Clare
School of Marine Sciences & Technology
Newcastle University
Ridley Building
Newcastle upon Tyne
NE1 7RU
United Kingdom

ISSN 1611-6119
ISBN 978-3-540-30014-4 Springer-Verlag Berlin Heidelberg New York

Library of Congress Control Number: 2005934892

Springer-Verlag is a part of Springer Science + Business Media
springer.com

Production: SPI Publishing Services, Pondicherry
Cover desing: design & production GmbH, Heidelberg, Germany

Printed on acid free paper 39/3150 YK 5 4 3 2 1 0

Preface to the Series

Recent developments in the applied field of natural products are impressive, and the speed of progress appears to be almost self-accelerating. The results emerging make it obvious that nature provides chemicals, secondary metabolites, of astonishing complexity. It is generally accepted that these natural products offer new potential for human therapy and biopolymer science. The major disciplines which have contributed, and increasingly contribute, to progress in the successful exploitation of this natural richness include molecular biology and cell biology, flanked by chemistry. The organisms of choice, useful for such exploitation, live in the marine environment. They have the longest evolutionary history during which they could develop strategies to fight successfully against invading organisms and to form large multicellular plants and animals in aqueous medium. The first multicellular organisms, the plants, appeared already 1000 million years ago (MYA), then the fungi emerged and, finally, animals developed (800 MYA).

Focusing on marine animals, the evolutionary oldest phyla, the Porifera, the Cnidaria and the Bryozoa, as sessile filter feeders, are exposed not only to a huge variety of commensal, but also toxic microorganisms, bacteria and fungi. In order to overcome these threats, they developed a panel of defense systems, for example, their immune system, which is closely related to those existing in higher metazoans, the Protostomia and Deuterostomia. In addition, due to this characteristic, they became outstandingly successful during evolution: they developed a chemical defense system which enabled them to fight in a specific manner against invaders. These chemicals are of low molecular weight and of non-proteinaceous nature. Due to the chemical complexity and the presence of asymmetrical atom centers in these compounds, a high diversity of compounds became theoretically possible. In a natural selective process, during evolution, only those compounds could survive which caused the most potent bioactivity and provided the most powerful protection for the host in which they were synthesized. This means that during evolution nature continuously modified the basic structures and their derivatives for optimal function. In principle, the approach used in combinatorial chemistry is the same, but turned out to be painful and only in few cases successful. In consequence, it is advisable to copy and exploit nature for these strategies to select for bioactive drugs. Besides the mentioned metazoan phyla, other animal phyla, such as the higher evolved animals, the mollusks or tunicates, or certain algal groups, also produce compounds for their chemical defense which are of interest scientifically and for potential application.

There is, however, one drawback. Usually, the amount of starting material used as a source for the extraction of most bioactive compounds found in marine organisms is minute and, hence, not sufficient for their further application in biomedicine. Furthermore, the constraints of the conventions for the protection of nature limit the commercial exploitation of novel compounds, since only a small number of organisms can be collected from the biotope. Consequently, exploitation must be sustainable, i.e., it should not endanger the equilibrium of the biota in a given ecosystem. However, the protection of biodiversity in nature, in general, and those organisms living in the marine environment, in particular, holds an inherent opportunity if this activity is based on genetic approaches. From the research on molecular biodiversity, benefits for human society emerge which are of obvious commercial value; the transfer of basic scientific achievements to applicable products is the task and the subject of *Marine Molecular Biotechnology*. This discipline uses modern molecular and cell biological techniques for the sustainable production of bioactive compounds and for the improvement of fermentation technologies in bioreactors.

Hence, marine molecular biotechnology is the discipline which strives to define and solve the problems regarding the sustainable exploitation of nature for human health and welfare, through the cooperation between scientists working in marine biology/molecular biology/microbiology and chemistry. Such collaboration is now going on successfully in several laboratories.

It is the aim of this new subset of thematically connected volumes within our series "Progress in Molecular and Subcellular Biology" to provide an actual forum for the exchange of ideas and expertise between colleagues working in this exciting field of "Marine Molecular Biotechnology". It also aims to disseminate the results to those researchers who are interested in the recent achievements in this area or are just curious to learn how science can help to exploit nature in a sustainable manner for human prosperity.

Werner E.G. Müller

Preface

Most benthic marine organisms have a pelagic dispersal phase in their life cycle. For larvae of sessile benthic marine invertebrates that have been studied most in this regard, the acquisition of developmental competence to settle, together with the prevailing hydrodynamics, determine to a large extent the duration of the larval phase. Settlement, at least for species that have been investigated in detail, is not random. Rather, larvae may respond in the water column to chemical cues emanating from the substratum, and/or upon contact to physicochemical and biological characteristics of the substratum and elect to settle or reject the surface. Some species have specific requirements for settlement sites. For these, a common finding is the ability to engage in complex searching behaviors on a surface in the face of hydrodynamic forces. Algal spores may also test surfaces prior to attaching permanently. If the surface that organisms settle on is artificial, the process is termed biofouling. Although this may be beneficial in the case of aquaculture species, e.g. colonization of ropes by mussels, more often biofouling is viewed as problematic. Major economic costs are associated with biofouling of e.g. ships' hulls, pipes of cooling systems for power plants, and conversely aquaculture operations when nets/cages are affected. Hull fouling is also a major vector for marine invasive species.

Considerable attention has focused on fouling of ships' hulls, which is, of course, an age-old problem. Until recently, the only solution to hull fouling was to kill the colonizing species, particularly with toxic metals. Organotins, such as tributyltin oxide (TBTO), incorporated into self-polishing copolymer coatings effectively controlled biofouling and were until recently in use by up to 70% of the world's fleet. TBT was shown, however, to negatively impact the environment, notably affecting non-target organisms. The best-known effects are imposex in neogastropods and shell thickening in oysters. Increasing concern over organotins lead to the International Maritime Organization (IMO) holding an international convention in October 2001. Parties to the convention agreed to ban the application of TBT-based paints from 1st January 2003, and to a total ban from 1st January 2008. It is unclear how effective this ban will be. As TBT-based antifoulants provided annual savings to the shipping industry estimated to be >$5 billion, alternatives are clearly needed. The major antifouling paint companies, recognizing that there is no future for TBT as an antifoulant, have introduced TBT-free self-polishing antifouling coatings as one solution. These coatings reportedly contain copper oxide as the major biocide with 'booster' biocide(s) to control copper-resistant species such as the green macroalgae of the genus *Ulva*. In turn, these biocides are attracting increasing attention for their possible deleterious impact on the environment. There is thus an urgent requirement for non-toxic or at least environmentally benign means to control fouling.

This need has prompted a number of major research programs on antifouling, which is now a major branch of marine biotechnology. This book aims to cover recent progress in this subject with a focus on chemical defenses employed by sessile marine organisms, some of which may involve microbial symbionts. Chapter 1 provides an overview of chemical defenses of marine invertebrates, while Chapters 2 to 5 deal with particular classes of marine natural product antifoulants, namely furanones from a marine red alga, isocyanoterpenoids and alkyl pyridiniums of sponge origin, and indole derivatives from a bryozoan. Chapter 6 reviews bacterial fouling; an understanding of eukaryote signaling and in particular prokaryote-eukaryote interactions may provide important clues to fouling control. Chapter 7 deals with a mechanism-based antifouling strategy, particularly from the viewpoint of larval signal transduction. Finally, chapter 8 describes 'state-of-the-art' natural product research and how to deal with nano/micro molar amounts of marine natural products. We acknowledge all the authors for their contributions and devotion to this project.

Finally, we are grateful to Professor Werner E. G. Mueller for the opportunity to contribute to this interesting book series and to Ursula Gramm for her patience and assistance during preparation of this book.

Nobuhiro Fusetani
Anthony S. Clare

March 2005

Contents

Defense of Benthic Invertebrates Against Surface Colonization by Larvae: A Chemical Arms Race

P.J. Krug

Abstract. Sessile invertebrates evolved in a competitive milieu where space is a limiting resource, setting off an arms race between adults that must maintain clean surfaces and larvae that must locate and attach to a suitable substratum. I review the evidence that invertebrates chemically deter or kill the propagules of fouling animals and protists under natural conditions, and that chemosensory mechanisms may allow larvae to detect and avoid settling on chemically protected organisms. The fouling process is an ecologically complex web of interactions between basibionts, surface-colonizing microbes, and fouling larvae, all mediated by chemical signaling. Host-specific bacterial communities are maintained by many invertebrates, and may inhibit fouling by chemical deterrence of larvae, or by preventing biofilm formation by inductive strains. Larval settlement naturally occurs in a turbulent environment, yet the effects of waterborne versus surface-adsorbed chemical defenses have not been compared in flow, limiting our understanding of how larvae respond to toxic surfaces in the field. The importance of evaluating alternative hypotheses such as mechanical and physical defense is discussed, as is the need for ecologically relevant bioassays that quantify effects on larval behavior and identify compounds likely to play a defensive role in situ.

1 Introduction

1.1 Multiple Levels of Antifouling Defense

Studies of chemical defense against fouling are generally conducted from the perspective of the chemist or benthic ecologist, striving to uncover the mechanisms that protect biological surfaces from epibiosis. Larvae of fouling organisms are often treated as an enemy to be defeated, rather than a highly adapted marvel to be understood in an ecological context. The

P.J. Krug
Department of Biological Sciences, California State University, Los Angeles, 5151 State University Dr., Los Angeles, CA 90032-8201

Progress in Molecular and Subcellular Biology
Subseries Marine Molecular Biotechnology
N. Fusetani, A.S. Clare (Eds.): Antifouling Compounds

complexity of the evaluations larvae make before and after contacting a surface is daunting, and given their simple neural wiring, illustrates how natural selection hones the machinery governing settlement behavior. I offer this review from the perspective of a larval biologist interested in chemically mediated behavior, and will examine antifouling defense less as a desired commercial outcome but rather a natural obstacle facing larvae that must recruit to a suitable microhabitat or die. Various mechanistic options will be considered, and the ecological and chemical evidence for each weighed. In particular, four levels of defense will be examined by asking:

(1) Do invertebrates produce antimicrobial compounds to reduce bacterial abundance on their surfaces, thus eliminating larval settlement cues associated with biofilms?
(2) Do invertebrates attract specific bacteria to their surfaces, which then (a) chemically inhibit competing, inductive bacteria, or (b) directly deter fouling larvae?
(3) Do invertebrates rely on lipophilic secondary metabolites to repel exploring larvae, or to kill recently settled larvae and juvenile stages?
(4) Do invertebrates release waterborne signals that trigger behavioral rejection by larvae prior to contact?

From the marine natural products literature, nearly a hundred compounds demonstrated toxicity towards larvae or antisettlement activity in laboratory bioassays. Antilarval and antisettlement compounds have been comprehensively reviewed at 7–8 year intervals as the field has developed; the reader is referred to the scholarly works of Davis et al. (1989), Clare (1996), and Fusetani (2004). Few compounds have been tested in field assays or in moving water, however, which is needed to evaluate the ecological role of a putative antifouling compound. This review will focus on cases where a chemical defense has been tested against multiple fouling organisms, or ideally against the full guild of potential epibionts in the field.

1.2 Overview of the Fouling Process

Most marine invertebrates and algae have a microscopic, dispersing stage in their life cycle. For sessile, benthic species, commencement of the adult stage requires successful colonization of a suitable surface. This produces intense competition for available space, with prokaryotes, protists, and animal larvae rapidly colonizing inanimate and undefended biological surfaces. Fouling can have severely deleterious effects on organisms, such as inhibition of photosynthesis, blockage of filter feeding, and elevated risk of mechanical dislodgement or predation. In consequence, the planktonic pool of free bacteria, algal spores and competent larvae imposes a strong selective pressure on benthic invertebrates.

Epibiosis and fouling are common life-history traits in marine organisms, in part because dissolved organic matter and particulate food are effectively distributed throughout the water column. This stands in contrast to terrestrial systems, where few nutrients are suspended in air. Suspension feeding is thus the dominant mode of heterotrophy among the invertebrates that comprise fouling communities. This represents a double-edged sword for sessile organisms; the same currents that supply their food also carry an inexhaustible supply of propagules searching for a surface to which they may attach. Thus, an evolutionary arms race among sessile invertebrates is established: as larvae, they must locate and colonize a surface in order to metamorphose; yet as adults they must keep their own surfaces clean and ward off settlement by larvae. Selective pressures on larvae produce mechanisms for locating, exploring, and attaching to available substrata, while selection on adults drives the evolution of antisettlement strategies.

Competition for space represents an ecological force comparable to predation, yet the field of chemical ecology has disproportionately focused on defenses against predators and herbivores rather than colonization of surfaces (Pawlik 1993; Hay 1996). In part, this stems from the different time scales at which predation and surface colonization occur. Predation is a rapid process, easy to observe and tractable to experimental manipulation; a predator either consumes a prey item, or it does not. This is not to trivialize the ecological complexity of predator–prey or herbivore–algae interactions, but to highlight differences that affect experimental design. Biofouling and epibiosis are multi-step sequences culminating in the establishment of a mature community composed of prokaryotes, fungi, protists and invertebrates. This intrinsically complex process results from the web of interactions in the initial biofilm and subsequent community of colonizers, and is not as easily studied in laboratory or field experiments as predation or competition between adults.

1.3 Biofouling as a Sequential Ecological Process

Biofouling has been described as a four-step sequential process (Wahl 1989). The first two steps, which produce a microbial biofilm, occur in a similar manner whether on a crab carapace in the sea or on a catheter in a hospital room. The next two steps are unique to aquatic habitats, which involve the attachment of unicellular and multicellular eukaryotes to an inorganic or living surface.

The initial step, adsorption of organic macromolecules, occurs almost immediately after submersion of any surface. This "conditioning" process coats surfaces with a film composed chiefly of proteins, glycoproteins and polysaccharides, to which bacteria subsequently attach. The exact nature of the conditioning film depends on surface characteristics of the substrate, and the adsorbed layer may be highly heterogeneous (Taylor et al. 1997; Compere et al. 2001). The second step comprises colonization by prokaryotes and the subsequent development of a bacterial biofilm, within an hour of surface immersion in water. Initial colonizers penetrate the viscous sublayer to contact the surface, either passively carried by eddies or via cellular motility by flagella or pili (O'Toole and Kolter 1998). Cells then make contact with the adsorbed layer of organics through non-covalent interactions with cell-surface carbohydrates or adhesive proteins, and finally through covalent bonds to the outer cell wall. Surfaces of different wettability may require alternative attachment strategies by bacteria, and may also determine the strength of biofilm adhesion (Baier 1981; Fletcher and McEldowney 1984; Paul and Jeffrey 1985).

Once attachment to the surface has occurred, bacterial cells begin producing a matrix of extracellular polymeric substances (EPS) that is critical for maintaining adhesion and subsequent biofilm development; the EPS is composed of polysaccharides, proteins, and even DNA (Sutherland 2001; Whitchurch et al. 2002; Allison 2003). The chemical nature of EPS is now recognized as a critical determinant of biofilm architecture, strength, and material properties (Hall-Stoodley and Stoodley 2002), but our understanding of the diversity and functional consequences of EPS from different biofilms is in its infancy (see also the chapter by Callow and Callow).

Recent advances in genomics, proteomics, and analytical chemistry have revealed the importance of cell–cell signaling and global regulatory networks in biofilm development. Biofilm formation is an interactive process affected by local hydrodynamics of the fluid environment, physicochemical properties of the surface, and behavioral responses of bacterial colonizers, which quickly change gene expression and phenotype upon attachment (Geesey 2001). Genomic studies of the medically relevant bacterium *Pseudomonas aeruginosa* found changes in regulation of genes associated with flagella or pili and polysaccharide biosynthesis during biofilm formation; proteomic comparisons found differential expression of proteins involved in primary metabolism in biofilm versus planktonic cells (Whitely et al. 2001; Sauer et al. 2002). Biofilm phenotypes change with time and composition of the bacterial community, and represent a growing area of investigation both ecologically and biomedically. Quorum-sensing pathways, which allow density-dependent changes in bacterial phenotype, modulate biofilm formation as well as virulence and symbiosis (Davies et al. 1998, and chapters by Callow and Callow, and de Nys et al.).

After establishment of the primary biofilm, secondary colonization by unicellular eukaryotes occurs; these include photosynthetic taxa such as diatoms, and heterotrophic suspension feeders and predators (Cuba and Blake 1983). This stage begins days after immersion, and continues as the microbial community coating the immersed surface develops. The final step is the attachment of propagules of multicellular organisms, invertebrate larvae and algal spores. This produces a community of macro-organisms that is subsequently shaped by ecological processes such as competition, predation, and succession. Fouling is an on-going process with no true end, as even a mature fouling community will undergo changes in composition due to season, disturbance, predation, and other biological and abiotic influences. Development and final structure of the community may be strongly subject to supply-side trends such as reproductive seasonality and large-scale oceanographic processes controlling the delivery of larvae (Roughgarden et al. 1988; Thomason et al. 2000).

Fouling progresses differently in distinct habitats, and on nearby yet distinct substrates. The predominant organisms differ in temperate zones and the tropics, as is often reflected in the bioassay organisms used in antifouling research. Primary fouling threats in temperate regions are barnacles and bivalves; in tropical areas tube-building polychaetes like *Hydroides* are a major component of early fouling communities, and at the poles, diatoms are key fouling organisms (Hadfield et al. 1994; Slattery et al. 1995). Selection may drive regional specialization of antifouling defenses, depending on the local assemblage of fouling organisms competing for space. However, different ecological processes can shape the community on spatially proximate surfaces; for instance, Keough (1984) found that recruitment explained the difference between fouling around benthic invertebrates versus on nearby docks.

1.4 Ecological Importance of Antifouling Defense Mechanisms

For any long-lived benthic organism, epibiosis must either be tolerated or combated due to the drawbacks associated with a colonized surface. Epibionts increase weight and drag, can reduce buoyancy, constrict range of motion, slow growth (Wahl 1996, 1997), and increase risk of detachment by flow (Dixon et al. 1981; Witman and Suchanek 1984). They can also facilitate manipulation of prey by predators such as crabs (Enderlein et al. 2003; Manning and Lindquist 2003). Shading can drastically diminish photosynthesis in algae (Bulthuis and Woelkerling 1983; Sand-Jensen and Revsbech 1987); epibiosis may comparably harm invertebrates that harbor photosynthetic symbionts, such as scleractinian corals and many sponges, but this has yet to be tested. Epibionts impede gas exchange and remove dissolved nutrients before they reach the host, and compete for food while interfering with feeding currents of basibionts (Wahl and Lafargue 1990).

Less considered are the potential benefits of a fouled surface. Epibionts can provide camouflage against predators hunting by visual or chemical cues (Fishlyn and Phillips 1980; Wicksten 1983). Shell fouling by sponges reduces predation risk for bivalves (Bloom 1975; Vance 1978; Forester 1979; Feifarek 1987; Laundien and Wahl 1999); these interactions may be mutualistic, with the sponge benefiting from the water currents or mobility of the overgrown bivalve. Spines in the thorny oyster *Spondylus americanus* did not deter predation, but rather increased fouling and directed the growth of sponges beyond the vulnerable shell margin (Feifarek 1987). Epibionts that are themselves chemically defended can confer protection on their host, and may be intentionally exploited for this purpose (Barkai and McQuaid 1988; Stachowicz and Hay 1999). The mutualistic nature of beneficial epibionts suggests that colonization may be preferentially induced by the basibionts; an improved understanding of such systems might be a valuable counterpoint to antifouling investigations.

2 The Role of Biofilms and their Constituent Microbes in Fouling

Biofilms have long been recognized as fundamental settlement cues for many invertebrate larvae (Crisp and Meadows 1963; Crisp 1974). Microbial films are particularly important cues for sessile species that colonize hard substrata, such as sponges (Woolacott and Hadfield 1996; Maldonado and Young 1999), cnidarians (Leitz and Wagner 1993; Negri et al. 2001), mollusks (Tamburri et al. 1992; Zhao and Qian 2002), tube-building polychaetes (Kirchman et al. 1982; Unabia and Hadfield 1999; Harder et al. 2002a), barnacles (Wieczorek et al. 1995), bryozoans (Mihm and Banta 1981; Brancato and Woollacott 1982; Keough and Raimondi 1995) and ascidians (Szewzyk et al. 1991; Wieczorek and Todd 1997). Bacterial films also provide positive and negative cues for the attachment of algal spores (Joint et al. 2000). However, interpreting the effects of films on the fouling process is complicated. The composition of films and the behavior of invertebrate larvae both change with age, meaning that small changes in experimental design can profoundly alter results. Further, laboratory and field experiments often yield conflicting data.

2.1 Interpreting the Effects of Biofilms on Larval Settlement

Early studies variously reported microbial induction or inhibition of larval settlement for different species. It is now recognized that the nature of biofilms varies widely, and can present a range of positive and negative stimuli to settling larvae (Keough and Raimondi 1996). Larvae can potentially extract much information about a surface from the microbial film coating it, including length of submergence (Wieczorek et al. 1995), tidal height (Strathmann et al. 1981; Thompson et al. 1998; Qian et al. 2003), and local hydrodynamics (Neal and Yule 1994b). The characteristics of a filmed surface can also be modified by other organisms; for instance, cyprid "footprints" left behind by early explorers are settlement cues for subsequent cyprids (Walker and Yule 1984; Clare and Matsumura 2000). Recent occupation by conspecifics (Thompson et al. 1998) can further affect response to a filmed substrate. Sensitivity to such a broad range of cues is clearly adaptive to settling larvae, but will likely confound lab-based assays that manipulate only one or two parameters at a time. Indeed, a pronouncement that films are inductive or inhibitory for a given species is unlikely to be true in all contexts, given the complex interactions between the age of larvae, age and source of the film, nature of the substratum, and other factors that determine a film's bioactivity.

As a case study, the effects of biofilms on barnacle settlement have a convoluted history. Early claims that films facilitated cyprid settlement (Crisp and Meadows 1963) were followed by studies showing inhibitory effects (Maki et al. 1988); however, the latter results have been questioned due to the large number of cyprids per assay dish (Wieczorek et al. 1995; Head et al. 2003) and statistical methods used in data analysis (Keough and Raimondi 1995). Recent studies have shown that many factors feed into the interpretation of biofilms by cyprids of *Balanus amphitrite*. In laboratory experiments, settlement was inhibited by young biofilms but induced by mature films, and cyprids discriminated among filmed surfaces from different tidal heights (Wieczorek et al. 1995; Thompson et al. 1998). A separate study found no effect of biofilm age, but did find an effect of cyprid age: young cyprids were inhibited by films whereas older cyprids were induced by a filmed surface (Harder et al. 2001). Thus, using either cyprids or biofilms of a fixed age will miss key interactions between these variables.

Field studies further complicate the picture, often conflicting with laboratory results. Thompson et al. (1998) found that cues from prior occupation by conspecifics increased cyprid settlement in both lab and field studies, while other factors made a surface attractive either in the laboratory (biofilm age) or in the field (proximity to adult barnacles). These apparent contradictions were attributed to a scaling issue, with adult presence influencing settlement at spatial scales greater than 3 cm and microbiota dictating attachment sites at smaller scales (such as come into play in limited-volume laboratory assays). Studies with other barnacle species have revealed different settlement patterns. For *Elminius modestus*, attachment strength of cyprids was inversely proportional to biofilm age (4-day-old vs. 1-month-old film) (Neal and Yule 1994a). Keough and Raimondi (1995) found that *Balanus variegatus* recruitment decreased with film age, while *E. modestus* was negatively affected by films of all ages.

Complex results have also been reported for arborescent bryozoans and solitary ascidians, major contributors to mature fouling communities that are commonly used in laboratory bioassays. In field experiments, recruitment of the bryozoans *Bugula neritina* and *B. stolonifera* was unaffected by the presence of a film in one trial, but increased with film age in a second trial (Keough and Raimondi 1995). Wieczorek and Todd (1997) reported that *Bugula flabellata* larvae were inhibited by biofilms ranging from 1 to 12 days old. Mihm and Banta (1981) found that a biofilm reversed larval preference in *B. neritina* for unfilmed plastic over glass, and this effect was not related to changes in surface wettability. Tadpole larvae of the ascidian *Ciona intestinalis* were induced to settle by biofilms and the effect increased with biofilm age; this was attributed to a combination of larval preference and passive entrapment on the biofilmed surface (Wieczorek and Todd 1997). However, Keough and Raimondi (1995) found no effect of biofilm presence or age on recruitment of four ascidian species, including *C. intestinalis*. These results emphasize the need for a thorough understanding of factors that contribute to habitat selection in the field, if laboratory data are to be meaningfully interpreted.

The cues from biofilms are thought to be primarily prokaryote-derived biochemical signals. However, microbial eukaryotes such as diatoms may also be important. For instance, larvae of the tube-building polychaete *Spirorbis borealis* preferentially settled on films of the diatom *Navicula* sp. or mixed diatom films, but avoided films of the unicellular green alga *Dunaliella galbana* (Meadows and Williams 1963). Diatom films also influence larval settlement for other polychaetes (Harder et al. 2002b; Lam et al. 2003), echinoderms (Ito and Kitamura 1997), and mollusks (Daume et al. 1999).

2.2 Specific Bacterial Strains within Biofilms as Positive Cues for Settlement of Fouling Larvae

Although biofilms are inherently multi-kingdom assemblages, much evidence now exists that individual bacterial strains can produce specific stimulatory or inhibitory chemical signals. Further, settlement-inducing activity does not appear to be phylogenetically constrained, as closely related strains vary widely in their effects. For instance, the tropical fouling polychaete *Hydroides elegans* preferentially colonizes biofilms at least 3 days old (Hadfield et al. 1994). Although settlement correlated with bacterial abundance in a natural microbial assemblage, 13 strains were isolated that were strongly inductive, and a further 11 were moderately active at triggering metamorphosis; larvae were induced to settle by low molecular weight water-soluble products of the active strains (Unabia and Hadfield 1999). Subsequent work revealed that of four biofilm species, two strains that were 30% divergent at the 16 S rRNA gene were the most inductive to larvae, whereas a third bacterium that was only 3% different from an inductive strain had no activity (Huang and Hadfield 2003). In parallel investigations, over half of 38 bacterial strains were inductive to *Hydroides* larvae, and again no phylogenetic pattern to the activity was evident (Lau et al. 2002); however, bacterial cues were most effective when adsorbed onto a solid substratum (Harder et al. 2002a). A marine *Pseudomonas* sp. strain S9 induced settlement of larvae of *Ciona intestinalis*, and the settlement-promoting activity was due to chemical features of the bacterial exopolysaccharide, which in turn was dependent on the metabolic state of the bacteria (Szewzyk et al. 1991).

3 Modulation of Surface Bacteria by Invertebrates: Direct and Indirect Effects on Fouling

Given that microbial films are a prerequisite for fouling by most larvae, one antifouling strategy is for an invertebrate to lower the abundance of settlement-inducing bacteria on its surface. I will consider three chemically mediated ways such a strategy could operate, with examples of each from the literature. The first is to diminish bacterial surface abundance, thereby removing an inductive cue for larvae. Three possible means to this end are (a) to produce antibacterial metabolites, (b) to chemically repel bacteria, or (c) to inhibit biofilm formation by interfering with bacterial signaling pathways. The second strategy is to maintain a host-specific bacterial community comprising non-inductive strains. This could work by selective chemical attraction of non-inductive strains, and/or repulsion of inductive strains by host or "host-friendly" bacteria. For the third mechanism, the host relies on surface-associated bacteria to chemically repel larvae or kill juveniles of fouling organisms.

3.1 Antibiotic Chemistry: Maintaining a Bacteria-Free Surface as a Defense against Fouling

Compounds that inhibit development of a surface biofilm could indirectly prevent fouling by removing this essential settlement cue for larvae. The antimicrobial activity of invertebrate secondary metabolites has been appreciated for decades (Burkholder and Rutzler 1969; Faulkner 1984). Sponges produce a wealth of compounds with antibiotic activity against terrestrial and marine bacteria (Bergquist and Bedford 1978; Amade and Chevolot 1982; Thompson et al. 1985; Newbold et al. 1999). In a survey of Caribbean sponges, 48% of species were antibacterial to at least one of eight assay strains, and 23% of all extract-strain interactions were inhibitory (Newbold et al. 1999). Similar patterns of widespread, broad-spectrum antimicrobial activity have been reported for gorgonians (Jensen et al. 1996; Koh et al. 2002). Antibiotics may be produced by bacterial symbionts, suggesting a role in interspecific competition (Unson and Faulkner 1993; Unson et al. 1994; Bewley et al. 1996); if a consequence of this competition is reduced biofouling or disease of the basibiont, then the stage is set for adaptive coevolution.

Although intuitively appealing, antimicrobial chemistry is not clearly correlated with bacteria-free surfaces. Studies with ascidians and sponges suggest that low bacterial surface counts result from repellent compounds that deter bacterial attachment, rather than from antibiotics. Surface abundance of bacteria in 11 temperate ascidians was negatively correlated with chemistry that deterred attachment, but not with antimicrobial activity (Wahl et al. 1994). For instance, bacteria were over two orders of magnitude more abundant on *Polyclinum planum* than on *Cystodytes lobatus*. In an innovative assay, extracts of *C. lobatus* were highly deterrent to bacterial colonization of agar plugs, an activity that correlated with surface counts; *C. lobatus* also contained mild antimicrobial activity, but this did not explain bacterial abundance. Deterrent and antibiotic activities were attributable to different molecules. A survey of 26 Caribbean sponges produced a similar finding: 81% of sponge extracts reduced bacterial attachment to less than 40% of controls (Kelly et al. 2003). Six of 13 sponge species showed potent inhibition of bacterial attachment, but no detectable antibiotic activity (Newbold et al. 1999; Kelly et al. 2003). These studies suggest that chemically mediated repulsion of bacteria may account for unfouled surfaces more than metabolites that inhibit growth or kill bacteria.

The strongest evidence for this strategy comes from algae that interfere with the cell–cell signaling processes that control biofilm formation. The red alga *Delisea pulchra* produces furanones, small molecules that interfere with the acyl homoserine lactone (AHL) signaling pathway used by many bacteria to cue swarming and biofilm formation (Maximilien et al. 1998; Steinberg et al. 1998, 2001, see also de Nys et al. chapter). Eukaryotic interference with AHL systems is known for vascular plants and marine algae, but not among animals; however, host-specific bacteria may be able to use chemical signals to interfere with AHL systems in competing strains (Bauer and Robinson 2002; Taylor et al. 2004a). Such antagonism between bacteria is considered next.

3.2
Maintaining a Community of Host-Specific Bacteria to Block Inductive Biofilm Formation

Both culturing and molecular studies have shown that many marine organisms maintain species-specific microbial communities upon their surfaces, or within their bodies, that are distinct from the bacterial populations in surrounding seawater or on nearby surfaces (Hentschel et al. 2002). A new phylum-level clade of bacteria, *Poribacteria*, may exclusively associate with sponges (Fiesler et al. 2004). A study of bacteria associated with three sponge species found that all species harbored sponge-specific strains, and each species had a subset of host-specific strains that were absent from other sponges and adjacent seawater (Taylor et al. 2004b). In a survey of hard corals, Rohwer et al. (2002) found over half of bacterial isolates from coral surfaces were new genera or species. Different corals had host-specific microbial communities shared by conspecifics separated widely in space, and maintained for over a year, yet that were distinct from bacteria on adjacent corals of other species (Rohwer et al. 2002). Less is known about the microbial associations of ascidians. A dominant photosynthetic symbiont was found on multiple didemnid ascidians, and its presence was negatively correlated with diversity and abundance of other bacteria (Wahl 1995); the host may provide a particularly favorable environment for this symbiont, and/or the symbiont may out-compete or chemically inhibit competitors.

Instead of itself producing antibacterial compounds, an alternative strategy is for the basibiont to attract select bacteria to its surface, which can then competitively exclude other biofilm-forming bacteria. Relying on interspecific aggression between microbes relieves pressure on the host to synthesize costly chemical defenses, while keeping its surface free from fouling by most bacteria and, by extension, macro-organisms. Support for this comes from studies of algal surfaces. Of 280 microbial isolates from

Scottish marine macroalgae, 21% inhibited growth of at least one out of nine marine fouling bacteria isolated from natural biofilms (Boyd et al. 1999). A subset of these strains was then screened for induction of negative chemotaxis in two strains of fouling bacteria, using a novel bioassay. Culture supernatants from algae-associated bacteria were incorporated into agar at the bottom of a spectrophotometer cuvette; a reduction in optical density indicated that bacteria moved away from the agar, and thus that the algae-associated strain released repellent metabolites. Out of 21 strains assayed, 38% triggered avoidance behavior in the fouling bacteria. With only a single exception, none of the strains that induced negative chemotaxis produced antibiotic metabolites in liquid culture (Boyd et al. 1999). However, subsequent studies showed that some algae-associated strains produce antimicrobial compounds when grown as a biofilm (Yan et al. 2002, 2003) or when exposed to potential competitor bacteria or their growth media (Mearns-Spragg et al. 1998).

These studies highlight the dual significance of antagonism between microorganisms and microbial chemotactic behavior in the battle for surface colonization. The data suggest a model in which select, sometimes species-specific, bacterial strains are attracted to a host basibiont's surface. Upon attachment they form a biofilm, and at some critical density begin production of antimicrobial compounds; they also release soluble repellents to trigger avoidance in competing strains of fouling bacteria. In this manner, symbiosis is established between the basibiont and its surface microbiota; the bacteria receive a place to grow in exchange for warding off strains that would otherwise promote fouling, to the detriment both of the host and its associated microbes (Armstrong et al. 2001). Algae that only assume their natural morphology when grown with bacteria may be obligately co-adapted to such mutualistic relationships (Takewati et al. 1983).

Marine bacteria cultured from seawater, sediment, and living or inert surfaces are a diverse source of bioactive chemistry (Fenical 1993; Renner et al. 1999). Bacteria producing antimicrobial and other bioactive substances have been isolated and/or characterized from a range of invertebrate surfaces, including sponges (Unson et al. 1994; Bewley et al. 1996), gorgonians (Tapiolas et al. 1991), other cnidarians (Trischman et al. 1994), molluscs (Armstrong et al. 2000) and echinoderms (Burgess et al. 1999). However, the importance of microbial defense against pathogens, predators and epibionts of host animals remains unclear (Engel et al. 2002). In few cases are both host specificity and the ecological role of a microbe-invertebrate association understood. Embryos of the shrimp *Palaemon macrodactylus* are protected from the pathogenic fungus *Lagenidium callinectes* by the egg-associated bacterium *Alteromonas* sp., which secretes the antifungal compound 2,3-indolinedione (1) (Gil-Turnes et al. 1989). Similarly, embryos of the American lobster (*Homarus americanus*) are coated in a monolayer of an unidentified bacterium that produces 2-(*p*-hydroxyphenyl)ethanol (Gil-Turnes and Fenical 1992). For a marine animal, regulating the bacteria on its surface may be an effective defense (or one component thereof) against fouling and overgrowth.

An intriguing reversal of this paradigm occurs in the ubiquitous fouling bryozoan *Bugula neritina*, which contains low levels of bryostatins, potent cytotoxic macrolides (Pettit et al. 1982). Adult colonies contain a bacterial endosymbiont, *Endobugula sertula*, and vertically transmit it to their lecithotrophic larvae (Davidson and Haygood 1999). Knock-out experiments with antibiotics suggest that the symbiont produces the bryostatins (Davidson et al. 2001). Ecologically, this symbiosis may primarily protect the large, short-lived larvae from predators (Lopanik

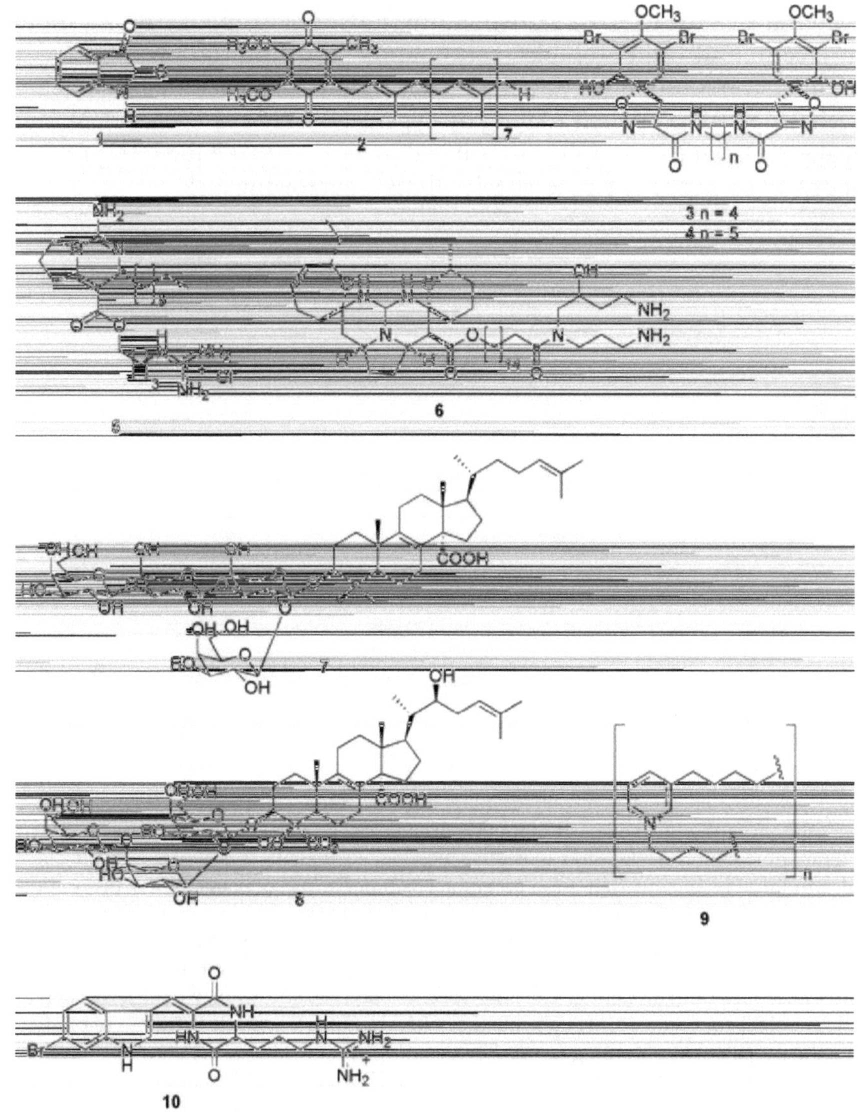

et al. 2004). Thus, microbe-larval associations may contribute to aspects of the fouling process beyond settlement induction, such as lowering larval mortality in the plankton.

3.3 Attracting Strains that Chemically Deter Settlement of Fouling Larvae

Just as specific bacterial strains may trigger settlement of certain larvae, others can inhibit surface colonization. Contact with three bacterial isolates inhibited settlement of *Balanus amphitrite* larvae, even in the presence of adult-derived positive cues (Lau et al. 2003). Five of six bacterial strains deterred settlement by *Bugula neritina* larvae, with both surface-bound and soluble compounds showing larval toxicity or settlement inhibition (Bryan et al. 1997). Numerous bacterial strains have been reported to prevent fouling by cyprids (Maki et al. 1988; Avelinmary et al. 1993), but the complications listed in Sect. 2.1 make the interpretation of cyprid response to single-species films a harrowing task.

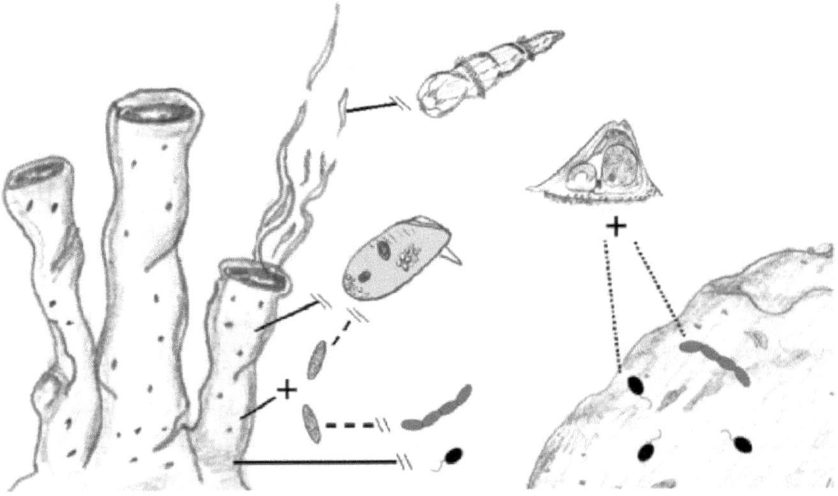

Fig. 1. Possible chemically mediated interactions between a benthic invertebrate, bacteria and other microbiota, and larvae of fouling organisms. *Pluses* denote positive, or attractive, interactions; *double hash marks* indicate repellency or toxicity. *Solid lines* indicate chemicals produced by the invertebrate, a potential basibiont at risk of being fouled. Invertebrates may produce non-polar metabolites concentrated along their surface, or polar compounds released into overlying water, possibly in waste water exiting through excurrent canals. The basibiont may attract species-specific, beneficial bacteria to its surface, while deterring other microbes from attaching. The host-specific bacterial strains may themselves deter or kill microbial competitors through chemistry (*dashed lines*). Excluded strains are predicted to induce settlement of fouling larvae when coating an undefended organism or inanimate surfaces (*dotted lines*). Either the host or beneficial bacteria colonizing its surface may produce compounds that deter larval settlement, either by triggering behavioral rejection of the surface, or through sub-lethal toxicity or post-settlement toxic effects on larvae

Given that some bacterial strains inhibit settlement, another possible defense is for an organism to preferentially attract such inhibitory bacteria to its surface (Fig. 1). Bacteria in biofilms may not just repel competing microbes, but could also produce metabolites that deter or kill fouling propagules (Holmström and Kjelleberg 1994). For example, one of five active bacterial strains isolated from the surface of the adult tunic of *Ciona intestinalis* produced cytotoxic, low-molecular weight metabolites that killed both *C. intestinalis* and *Balanus amphitrite* larvae, and a less active, high molecular weight fraction that killed *B. amphitrite* cyprids (Holmström et al. 1992; Holmström and Kjelleberg 1993). A marine *Alteromonas* strain isolated from the sponge *Halichondria okadai* produced ubiquinone-8 (2), which inhibited barnacle cyprid settlement (Kon-ya et al. 1995). Bacterial isolates from the marine alga *Ulva reticulata* were non-toxic to five *Vibrio* strains and to larvae of *Hydroides elegans*; however, settlement of *H. elegans* larvae was inhibited by a water-soluble compound from *Vibrio* sp. two and by a biofilm of *Pseudoalteromonas* sp. two (Dobretsov and Qian 2002). Microbial metabolites were inhibitory even in the presence of the artificial inducer 3-isobutyl-1-methyxanthine (IBMX), suggesting competition for binding sites.

Bacterial strains associated with algal and invertebrate surfaces may be more likely to produce antifouling compounds than free-living strains. Holmström et al. (1996) compared the effects of microbial isolates from rock surfaces with those from algal and animal surfaces against *Balanus amphitrite* larvae, spores of the green alga *Ulva lactuca*, and marine diatoms. Of 93 isolates from rock surfaces, less than 10% blocked larval settlement, fewer than 20% inhibited algal spores from growing, and none of 10 strains affected diatoms; in contrast, 74% of algal isolates and 30% of animal isolates inhibited settlement of barnacle larvae (Holmström et al. 1996). Marine *Pseudoalteromonas* species from invertebrate and algal surfaces have received considerable attention for their potential significance as mediators of antifouling defense (Holmström and Kjelleberg 1999, 2000; Egan et al. 2000; Holmström et al. 2002).

Further support for the hypothesis that bacteria provide antifouling defenses for their invertebrate host comes from the soft coral *Dendronepthya* sp., which is rarely fouled in the field and has a characteristic community of surface bacteria. Chemicals from both the coral and surface-associated bacteria inhibited attachment and growth of other bacteria from nearby abiotic surfaces (Harder et al. 2003). Dobretsov and Qian (2004) isolated 11 bacterial strains that were unaffected by coral antibiotics, of which three inhibited attachment and growth of other marine fouling bacteria. Two strains were subsequently found to inhibit settlement of *Hydroides elegans* larvae, and generated an overall inhibitory effect when mixed with other, inductive bacteria into a biofilm (Dobretsov and Qian 2004). Waterborne, high-molecular weight polysaccharides produced by the two strains inhibited larval settlement of *H. elegans* and *Bugula neritina*. Settlement was blocked in the presence of the inducer IBMX, yet larvae remained viable and settled when transferred to clean seawater; thus, the water-soluble settlement deterrents were not toxic. Soluble carbohydrates may compete for binding sites on larval lectins, as in other systems (Kirchman et al. 1982; Bahamondes-Rojas and Dherbomez 1990).

Similar results were obtained with the sponge *Mycale adherens*, a member of tropical fouling communities that is not itself fouled. The sponge surface yielded only half the number of bacterial isolates as reference dishes submerged nearby for 5 days, and only 3 of 40 sponge isolates were also found on the reference surfaces; multiple genera of bacteria were found only on the sponge, suggesting a restricted community of microorganisms (Lee and Qian 2003). Whereas 65% of natural biofilm strains induced settlement of *Hydroides elegans* larvae, 75% of sponge-associated strains were non-inductive, and 40% inhibited larval settlement. Dissolved compounds from the sponge were toxic to larvae, but only inhibited the growth of inductive bacteria at 10-fold elevated concentrations.

Taken together, these data suggest a simple hypothesis: bacteria that *repel* fouling larvae are likely to be *attracted* to unfouled invertebrates, whereas bacteria that *induce* settlement in fouling larvae are likely to be *repelled* by the surfaces of unfouled animals. Although chemically mediated repulsion has been documented, attraction of bacteria to the host surface has received little attention for invertebrates. Given the high degree of specialization between marine organisms and their microbial communities, there is reason to expect such chemotactic interactions. Conversely, settlement-inducing strains may be chemically repelled by the basibiont itself (Wahl et al. 1994), or by host-specific bacterial colonists (Armstrong et al. 2001). Thus, there is potential for multi-tiered interactions: animals use chemistry to attract or repel surface-colonizing bacteria; bacteria produce chemicals that induce or deter settlement by other bacteria, algal spores or invertebrates; and the host organism can itself produce chemicals that induce or deter larval settlement (Fig. 1).

Such associations between invertebrates, bacteria, and larvae of epibionts are similar in their ecological complexity to the tri-trophic interactions between plants, herbivorous insects and parasitic wasps (Baldwin and Preston 1999). In these interactions, chemical signals form when caterpillar saliva contacts plant membrane lipids, inducing the attacked plants to produce volatile organic signals; these airborne molecules attract parasitic wasps that paralyze the caterpillars and lay eggs in them. Remarkably, signals between the insect's victim (plant) and parasite (wasp) can be species-specific, with the plant selectively attracting the wasp species that specializes on the type of caterpillar chewing that plant (de Moraes et al. 1998). Similarly complex chemical communication may occur between invertebrates, surface microbes, and fouling organisms, but detailed investigations are needed to define the scope and generality of these interactions.

4 Role of Basibont-Derived Chemistry in Defense against Eukaryotic Propagules

4.1 Sponges

Many studies have partially or fully characterized sponge metabolites that inhibit settlement, anaesthetize or kill larvae in laboratory still-water assays (Sears et al. 1990). However, with no knowledge of where these compounds are located within the living sponge, or at what concentrations they are present in water overlying a sponge, the ecological significance of antilarval activity is obscured. For instance, a series of non-polar terpenes and steroids from three sponges inhibited cyprid settlement without apparent toxicity; however, as each of these metabolites was cytotoxic or inhibited a key enzyme, their activity in laboratory assays may not reflect their role in nature (Tsukamoto et al. 1997). I will therefore focus on results from studies that tested ecologically relevant aspects of chemical defense against fouling organisms.

In one of the earliest and most cited papers to examine exudation of sponge metabolites, Walker et al. (1985) quantified release rates for the brominated compounds aerothionin (3) and homoaerothionin (4) from the intertidal sponge *Aplysina fistularis*. Rates of exudation were measured (1) in aquaria, using transplanted sponges, and (2) in situ by immersing exposed sponges in containers holding artificial seawater. This study was an admirable first step towards quantifying the release of potential antifouling compounds from a marine invertebrate, and it is striking that few comparable studies have been undertaken in the subsequent quarter-century. The study suffered for lack of replication, as only two sponges were successfully used for in situ work and six for aquaria experiments (Walker et al. 1985). This likely accounted for the high variance between replicates, making the measurements of exudation qualitative rather than quantitative. Release of aerothionin was measured as ~ 9 ng min^{-1} g^{-1} dry wt of sponge in aquaria, versus 0.8 ng min^{-1} g^{-1} dry wt in situ, further complicating the interpretation of these results. As injured sponges released up to 100 times more aerothionin than uninjured sponges, the higher release rates measured in aquaria may have resulted from stress or damage to sponges during transportation. However, it is clear that these metabolites are released from minimally disturbed sponges into the surrounding seawater at detectable levels.

The importance of aerothionin exudation was supported by ecological evidence and energy dispersive X-ray microanalysis of *Aplysina fistularis* cells (Thompson et al. 1983; Thompson 1985). The brominated metabolites were localized to spherulous cells, which are concentrated just under the exopinacoderm (the peripheral layer), or beneath the lining of excurrent canals that carry water out of the sponge (Thompson et al. 1983). Electron microscopy showed some spherulous cells degenerating such that their enclosed metabolites would be released into the mesohyl, or liberated into water flowing out of the sponge.

Whether these compounds are responsible for the observed antifouling properties of *Aplysina fistularis* remains unclear. Neither aerothionin nor a mixture of both metabolites prevented germination of brown algal spores of the kelp *Macrocystis pyrifera*, but this is not a fouling organism (Thompson et al. 1985). Neither compound prevented settlement in long-term assays with larvae of the bryozoan *Phidolophora pacifica* and the polychaete *Salmacina tribranchiata*; although these are more ecologically relevant organisms for fouling studies, the laboratory assays were run for exceptionally long periods (10 days and 7 days, respectively). Both aerothionin and the mixture of both metabolites stopped larvae of the abalone *Haliotis rufescens* from completing metamorphosis in the presence of the artificial inducer GABA (Thompson et al. 1985). Given that pesticides also block settlement of GABA-induced *Haliotis* larvae (Morse et al. 1979), the brominated compounds from *Aplysina fistularis* may simply be mildly toxic to abalone larvae, preventing settlement in short-duration (1 h) assays. Indeed, both compounds were toxic to brine shrimp larvae. These metabolites exhibited potent antimicrobial activity, however, which was correlated with reduced surface fouling in a survey of 40 Californian sponges (Thompson et al. 1985). The sponge may thus reduce fouling by slow release of antibiotics from rupturing spherulous cells, preventing growth of a primary biofilm.

Other studies have also indicated that bioactivity or specific metabolites can be concentrated in spherulous cells of sponges. In the Mediterranean sponge *Crambe crambe*, spherulous cells contained most of the toxicity found in whole-sponge extracts, due to alkaloids such as crambescin A (**5**) and crambescidins (e.g., **6**) (Uriz et al. 1996). The defensive chemistry of this sponge inhibited settlement of *Bugula neritina* larvae, and limited the growth of marine bacteria (Beccero et al. 1997). Spherulous cells were concentrated in the exopinacoderm of the sponge, and were microscopically observed exiting the sponge surface. The enclosed compounds are thus likely to be concentrated in, and released onto, the sponge surface, supporting their putative ecological role in antifouling.

The recent study by Kubanek et al. (2002) is an excellent model for how to comprehensively assess the ecological roles of marine natural products. In this investigation, triterpene glycosides from two sponges, *Erylus formosus* and *Ectyoplasia ferox*, were tested for a suite of activities, including predator deterrence, bacterial attachment, fouling, and overgrowth by competitors. Metabolites of *E. formosus* (e.g., formoside, **7**) collected from four different sites were effective antifeedants, and inhibited attachment of a *Vibrio harveyi* strain to agar blocks at concentrations well below those occurring in the sponge. Antifouling activity was assessed in field assays, with compounds incorporated into Phytogel; compounds were initially incorporated into gels at two times whole-sponge concentrations, but as gels lose ~50% of their content over a 3-week deployment, this approximated natural levels in the sponge for the length of the trial. Triterpene glycosides from *E. formosus* strongly inhibited fouling by invertebrates and algae over a 27-day period. Further, at concentrations measured only in the surface layer of the sponge, the compound formoside (7) inhibited fouling for 14 days (Kubanek et al. 2002). Structurally similar triterpene glycosides from *E. ferox* such as ectyoplaside A (**8**) were not active in antifouling and bacterial attachment assays, but were antifeedant to fish and allelopathic to another sponge. Given the effects against bacterial attachment, triterpene glycosides might inhibit fouling indirectly by preventing establishment of a biofilm; alternatively, the metabolites could act directly on settling larvae or spores, but this was not tested.

Two methods were used in an attempt to quantify levels of glycosides in seawater around sponges, but neither method was successful, either due to low release rates or to problems with co-eluting contaminants and low recovery efficiency of the sampling apparatus. Concentrations of the active compounds were lowest (2.1 mg ml^{-1} sponge) in the outermost layer of *Erylus formosus*, but were otherwise uniformly distributed through the remaining sponge (6.3–7.5 mg ml^{-1} sponge); surface swabs yielded an estimate of 0.04 mg/cm^2sponge. Although concentrations were low in the outermost layer, these levels were sufficient to inhibit fouling in a realistic field assay. In contrast, the antifeedant metabolites of *Ectyoplasia ferox* were most concentrated in the outermost 2 mm of the sponge. It is striking that the glycosides of *E. formosus* were antifouling, yet were more concentrated within the sponge, whereas those of *E. ferox* lacked antifouling activity but were present in the outermost layer of the sponge. This illustrates the importance of quantifying where metabolites are present within an invertebrate, as natural distributions may be counter-intuitive to our naïve predictions. Although it is frequently postulated that antifouling metabolites should be concentrated in surface layers of an animal, the study by Kubanek et al. (2002) demonstrates that natural selection does not necessarily produce such expected patterns.

The effects of antifouling sponge exudates were tested in the field by Dobretsov et al. (2004) by placing settlement dishes at varying distances from the unfouled sponge *Callyspongia pulvinata*, and from physically similar sponge mimics as controls. In laboratory assays, sponge compounds inhibited diatom growth and settlement of the polychaete *Hydroides elegans*, but were not antimicrobial. Analogously, sponges did not affect bacterial density or the settlement-inducing effects of microbial biofilms that grew on nearby plates in the field. Sponge proximity was negatively correlated with abundance of macroalgae and fouling invertebrates on nearby abiotic surfaces, however, supporting the hypothesis that the sponge's chemistry is responsible for keeping its surface clean. The antifouling strategy of *Callyspongia pulvinata* thus appears to be deterrence of spores and larvae via release of polar metabolites. In contrast, *Aplysina fistularis* may rely on the antimicrobial effects of compounds present at the surface or released into the environment to prevent biofilm development, and thus to inhibit subsequent colonization by larvae.

4.1.1 Potential Non-Toxic Antifoulants, Suggested by Laboratory Bioactivity

Although untested in field assays, some non-toxic sponge compounds inhibit settlement of fouling larvae, suggesting an ecological role. Polymeric 3-alkylpyridinium salts (poly-APS, **9**; see Sepčić and Turk chapter) from the sponge *Reniera sarai* had antisettlement effects on barnacle cyprids that were non-toxic and reversible (Faimali et al. 2003). These compounds have surfactant properties, forming macromolecular assemblages in aqueous solution; they are partly water-soluble, due to their de-localized charge, yet also form a "greasy layer" on the sponge surface, due to alkyl side chains. Poly-APS inhibited settlement of *Balanus amphitrite* cyprids at 1 μg ml^{-1}, a concentration that did not affect naupliar swimming or survival; further, the effects were fully reversible after 3 days of exposure to 10-fold higher levels (Faimali et al. 2003). In contrast, heavy metal antifoulants were toxic to nauplii, bivalve larvae and microalgae at low concentrations, equal in magnitude to their EC_{50}'s for settlement inhibition. Some marine natural products may thus deter settlement in fundamentally different ways from toxic metals.

The brominated diketopiperazine barretin [(6-bromo-8-entryptophan) arginine, **10**], from the sponge *Geodia barretti*, inhibited barnacle cyprid settlement with an EC_{50} of 0.4 μg ml^{-1} (Sjögren et al. 2004); the effects were non-toxic at 10-fold higher doses, and were fully reversed upon transfer to clean seawater. Reduction of the double bond in 8,9-dihydrobarettin caused a 10-fold loss of activity. In still water, *Geodia barretti* released sufficient barretin to inhibit settlement after a 10-fold dilution. This would be an ideal organism for future in situ measurements of exudation rates. Non-toxic brominated settlement inhibitors from other sponges include mauritiamine (**11**, EC_{50} = 15 μg ml^{-1}) and oroidin (**12**, EC_{50} = 19 μg ml^{-1}) from *Agelas mauritiana* (Tsukamoto et al. 1996a), ceratinamide A (**13**, EC_{50} = 0.1 μg ml^{-1}) from *Pseudoceratina purpurea* (Tsukamoto et al. 1996b), and a mildly active bromopyrrole (**14**, EC_{50} = 21 μg ml^{-1}) from *Agelas conifera* (Keifer et al. 1991).

Numerous terpenes with antisettlement activity against cyprids have been isolated from sponges like *Axinyssa* spp. and *Acanthella cavernosa*, and *Phyllidia* nudibranchs that consume them (Okino et al. 1995, 1996a,b;

Hirota et al. 1996, 1998). Potent non-toxic settlement inhibition (EC_{50}'s <1 μg ml^{-1}) is associated with isocyano functional groups, which may render molecules more inhibitory than $CuSO_4$(EC_{50}'s <0.15 μg ml^{-1}) without the associated toxicity. These compounds should be evaluated in the field using slow-release gels, and their exudation from sponges measured in situ.

The role of sterols in antifouling is unclear. Cholesterol endoperoxide (**15**) and the dieterpene manoöl (**16**) from the sponge *Aplysilla glacialis* enhanced the biomass of fouling organisms when painted onto a surface and deployed in the field (Bobzin and Faulkner 1992). Cholesterol inhibited settlement of ascidian larvae in laboratory assays, however (Davis et al. 1991). Neither cholesterol nor ergosterol were effective against cyprids, but halistanol sulfate salts were modestly antisettlement towards cyprids (EC_{50}'s 3–4 μg ml^{-1}) (Tsukamoto et al. 1997); further, steroids with a D-seco moiety from the octocoral *Dendronepthya* sp. inhibited cyprid settlement for 7 days, with no toxicity at elevated doses (Tomono et al. 1999).

4.2 Cnidarians

4.2.1 Cnidarian Defense against Microorganisms and Algae

Many cnidarians chemically inhibit diatom growth, possibly through non-toxic mechanisms (Wilsanand et al. 2001). Homarine (**17**, *N*-methyl-2-carboxy-pyridine) was isolated from the gorgonians *Leptogorgia virgulata* (0.3% wet weight) and *Leptogorgia setacea* (0.25% wet weight) (Targett et al. 1983). Both water-soluble extracts of the corals and the in situ concentration of homarine (2 mg ml^{-1}) inhibited growth of the co-occurring benthic diatom *Navicula salinicola* by 41%. Although not present in *Leptogorgia* spp., the related compounds nicotinic and picolinic acid were even more inhibitory. Homarine is common in the tissues of many invertebrates (Carr et al. 1996; Polychronopoulos et al. 2001), and may be involved in osmoregulation (Shinagawa et al. 1995), pattern formation during cnidarian development (Berking 1987), and donation of methyl groups in crustacean biosynthesis (Netherton and Gurin 1982). Unlike many nitrogenous polar metabolites, homarine does not stimulate fish feeding (Carr et al. 1996), and deterred predation by an Antarctic asteroid (McClintock et al. 1994).

Bandurraga and Fenical (1985) compared the unfouled octocoral *Muricea fruticosa* with its sympatric congener *M. californica*, which is commonly overgrown by fouling organisms. The unfouled *M. fruticosa* contained the muricins (e.g., **18**), saponins with differently modified aminogalactose residues linked to a degraded sterol derived from pregnane. Although none were active in cytotoxicity or antimicrobial assays, all four compounds inhibited growth of the diatom *Phaeodactylum* *tricornutum* by 60–80% at 100 ppm and completely abolished growth at 200 ppm. The basis for inhibition of diatom growth by homarine and the muricins is unknown.

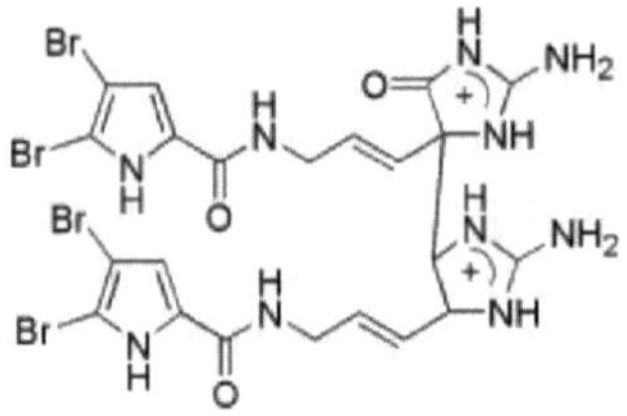

11

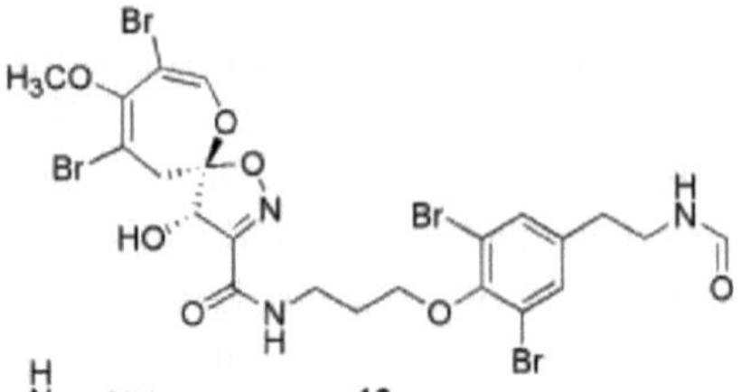

13

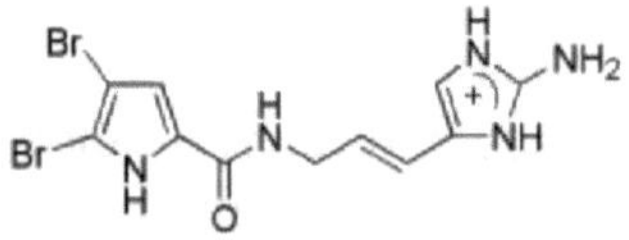

12

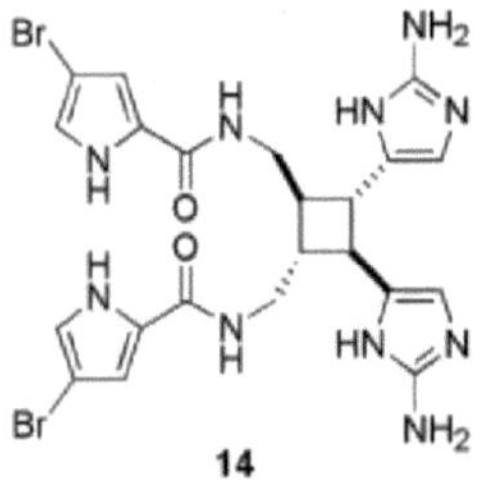

14

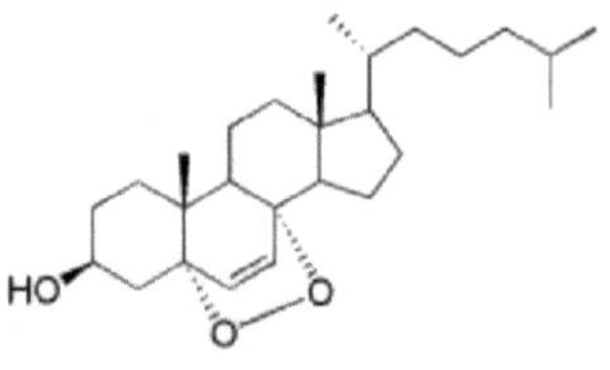

15

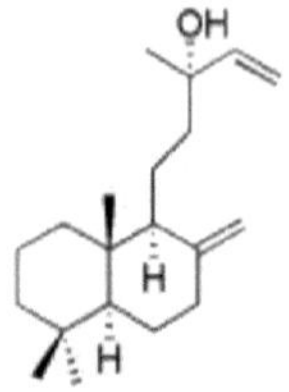

16

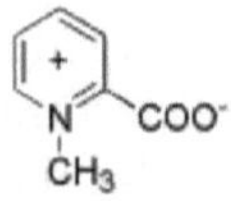

17

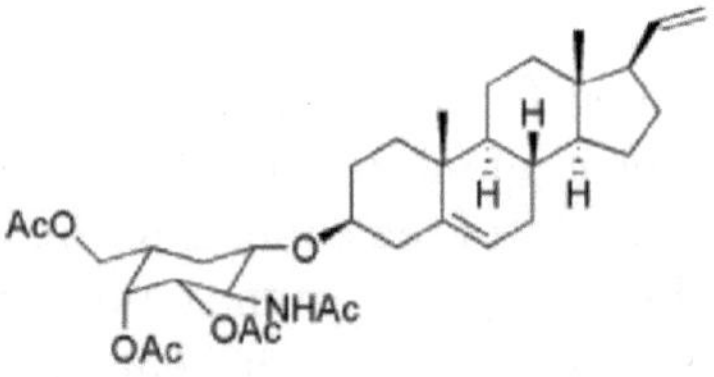

18

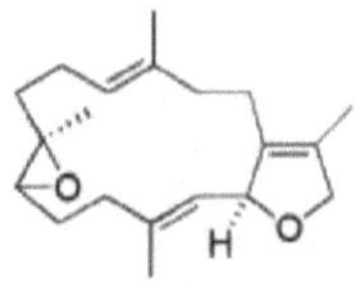

19

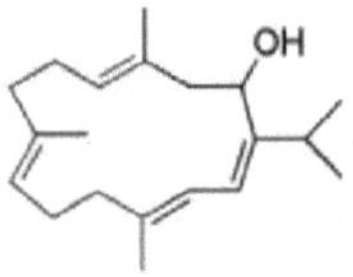

20

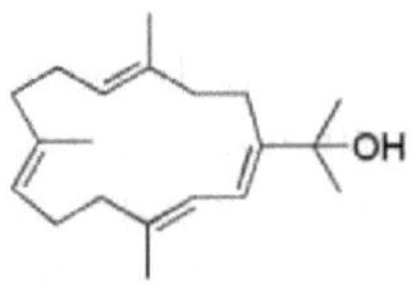

21

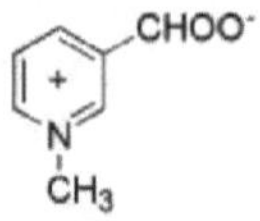

22

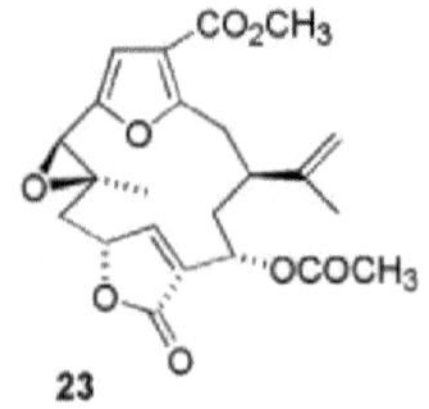

23

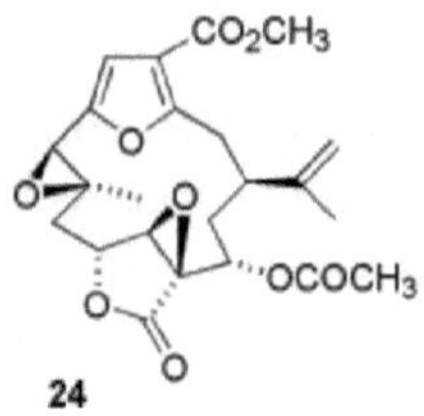

24

Targett et al. (1983) suggested that soft corals employ a two-level defense against fouling: chemicals slow the growth of photosynthetic protists, and periodic sloughing of the external covering removes the accumulated fouling community. A combination of physical and chemical defenses may operate in many invertebrates, and illustrates the need to consider alternative hypotheses in chemical ecology. For instance, Coll et al. (1987) reported differential patterns of algal fouling among colonies of the soft coral *Lobophylum pauciflorum* that corresponded with distinct chemotypes. Colonies overgrown by the algae *Ceramium flaccidum* and *Enteromorpha* sp. contained diterpene **19** with dihydrofuran and epoxide moieties; structure-activity studies implied that both oxygen functionalities were required for optimal bioactivity. In contrast, co-occurring colonies with clean surfaces contained a pair of cembrane diterpene alcohols **20–21**. Assays with *Ceramium codii* (a congener of the fouling alga) yielded the counterintuitive result that overgrown colonies contained **19**, which strongly inhibited algal growth, whereas clean-surfaced colonies produced **20–21**, the least inhibitory compounds out of nine tested.

The authors inferred that differences in chemical composition *resulted* from algal overgrowth, postulating that the less-inhibitory compounds were converted to the more active molecule in response to fouling. Clearly, however, if the clean colonies had contained the more-active metabolite, the inference would have been that the diterpenes prevented algal overgrowth. This illustrates the danger in drawing adaptationist conclusions from limited data, since two contradictory conclusions cannot both be right; either the chemical differences play a causative role in observed patterns of fouling, or they don't. There is a tendency in the marine chemical ecology literature to hypothesize causative relationships between an ecological phenomenon and the presence of chemistry, without subsequently testing the hypothesis with appropriate experiments. For instance, in the case of *L. pauciflorum*, the hypothesis that soft coral chemistry changes as an induced response to algal epibiosis could be tested by experimental manipulation of algal cover and subsequent studies of chemical composition. In fact, a survey of 20 soft coral species found no chemical differences between conspecific pairs of overgrown and clean colonies, and colonies were observed to regularly shed their algal epibionts. Chemical variation in soft corals may not explain algal fouling, and physical defenses may be critical to keeping cnidarian surfaces clean.

Slattery et al. (1995) compared two Antarctic soft corals with unfouled surfaces (*Alcyonium paessleri* and *Gersemia antarctica*) to one frequently fouled by diatoms (*Clavularia frankliniana*). Polar extracts of *A. paessleri* and *G. antarctica* had weak antimicrobial activity against sympatric marine bacteria, but chloroform fractions had strong antiattachment activity, showing that different compounds kill bacteria or inhibit attachment (e.g., Wahl et al. 1994). Both polar and non-polar fractions of *A. paessleri* and *G. antarctica* inhibited growth of the diatom *Navicula* sp. Aqueous homogenates, polar and non-polar fractions all impaired embryonic development of the sea urchin *Sterechinus neumayeri* in 48-h assays, but the relevance of this activity to fouling is unclear. Extracts of *C. frankliniana* had no bioactivity. Field assays were inconclusive; long-term trials with agar gels were not interpretable due to heavy overgrowth on treatment and control gels, and polar extracts quickly diffused from glass surfaces. Chloroform extracts of the unfouled species *A. paessleri* and *G. antarctica* limited diatom growth on treated glass slides over 1 month field deployments (Slattery et al. 1995). Although present in all three soft corals, homarine was not quantified in tested extracts (Slattery 1994). A subsequent study examined the biological activity of water-soluble compounds released by soft coral colonies maintained in aquaria. Water within 1–2 cm of *A. paessleri* contained a mixture of cholesterol and related sterols at a concentration of 1.54 ± 1.09 mg l^{-1}, while water surrounding *G. antarctica* contained principally homarine and its isomer trigonelline (**22**) at 1.26 ± 0.98 mg l^{-1} (Slattery et al. 1997). The polar exudate from *G. antarctica* inhibited growth of 3 bacterial strains, as did pure homarine, but the sterols from *A. paessleri* were not antimicrobial. Trigonelline was also isolated as an antisettlement agent from the octocoral *Dendronepthya* sp. (Kawamata et al. 1994).

4.2.2
Cnidarian Defense against Invertebrate Larvae

A low-molecular weight inhibitor of barnacle cyprid settlement was liberated into seawater by *Leptogorgia virgulata* (Standing et al. 1984), the same organism in which homarine slows diatom growth. However, high-molecular weight fractions induced settlement, and the bioassays were not ecologically interpretable. Inhibition was subsequently attributed to the diterpenes pukalide (**23**) and epoxypukalide (**24**) (Gerhart et al. 1988). Similar activity against *Balanus amphitrite* cyprids was reported for the renillafoulins (e.g., **25**), diterpenes from the octocoral *Renilla reniformis* (Keifer et al. 1986). Small analogues of these natural products containing furan or lactone rings inhibited cyprid settlement in a reversible manner not due to sub-lethal toxicity (Clare et al. 1999). More studies of how the parent natural products affect larval behavior would clarify the role of diterpenes in cnidarian antifouling defense. For instance, the renillafoulins inhibited settlement of cyprids but not of *Bugula neritina* larvae, and conversely, uncharacterized compounds inhibited *Bugula neritina* larvae but not cyprids, suggesting multiple compounds may be necessary for an effective non-toxic defense against a natural assemblage of fouling taxa (Rittschof et al. 1988).

Soft and hard corals may chemically affect the larvae of other corals, interactions at the interface of antifouling and allelopathy. Scleractinian coral recruitment was depressed in a current-dependent directional manner around the soft corals *Sinularia flexibilis* and *Sarcophyton glaucum*, and settlement did not occur on plates containing an extract of *S. flexibilis* (Maida et al. 1995). The diterpenes flexibilide (**26**), dihydro-flexibilide (**27**), and sinulariolide (**28**) from *S. flexibilis* were toxic to fertilized eggs of the hard corals *Montipora digitata* and *Acropora tenuis* during the first 24 hr, but not to sperm, unfertilized eggs, or planulae larvae (Aceret et al. 1995). The ecological significance of toxicity to embryos is unclear, as embryos are neither found near benthic corals nor naturally bathed in coral metabolites for 24 h. Koh and Sweatman (2000) reported that indole alkaloids from the hard coral *Tubastraea faulkneri* were toxic to planulae of 11 other coral species but not to conspecific larvae; however, similar caveats apply, since swimming larvae will not be continuously exposed to solutions of coral metabolites for 4 h prior to settlement. To determine how non-polar metabolites influence settlement by potential competitors, behavior of planulae should be quantified around surfaces exuding diterpenes. For instance, extracts of four hard corals caused reversible changes in larval behavior of the scleractinian *Pocillopora damicornis*, and slowed growth or killed newly settled juveniles (Fearon and Cameron 1997). Such data are necessary to evaluate the role of chemical defense against larvae of fouling organisms and potential competitors for space.

4.3 Ascidians

As a phylogenetic generalization, solitary ascidians are more heavily fouled than colonial ascidians; an exception is the solitary genus *Phallusia*, which was less fouled than co-occurring solitary ascidians in Bermuda, the Mediterranean, and southeastern Australia (Stoecker 1980; Uriz et al. 1991; Davis and White 1994; Bryan et al. 2003). In contrast, most colonial ascidians are extremely well defended against fouling. In Bermuda, 20 of 27 colonial tunicates were free of epibionts (Stoecker 1980); in the Mediterranean 17 of 18 species were unfouled (Uriz et al. 1991); and three of four colonial ascidians from Australia were devoid of fouling (Davis and White 1994). The tunic of solitary ascidians may be a favorable surface for settling larvae of epibionts, but no study has explicitly determined whether solitary and colonial ascidians differ in their degree of chemical protection against fouling larvae.

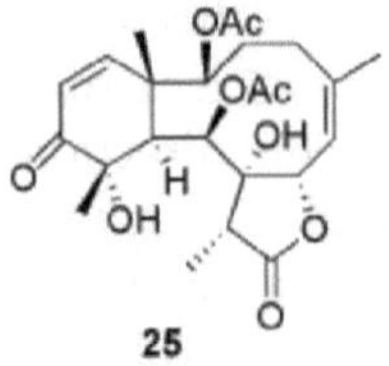

25

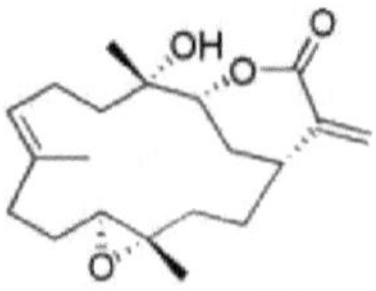

26

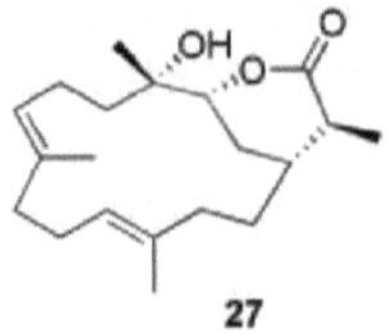

27

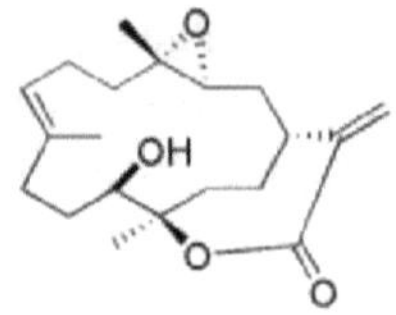

28

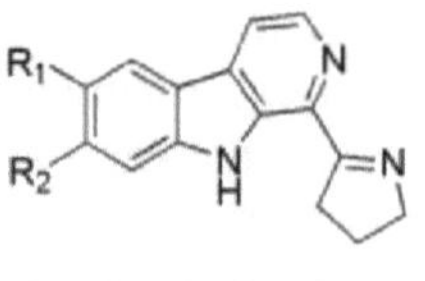

29 R_1 = H, R_2 = Br
30 R_1 = Br, R_2 = H

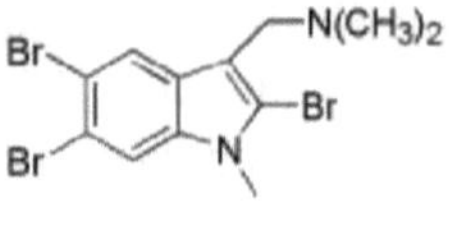

31

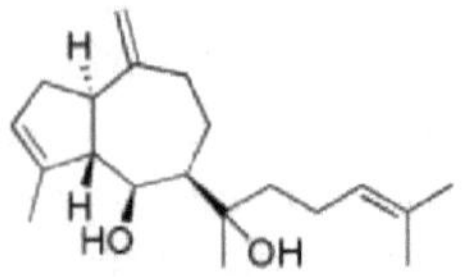

32

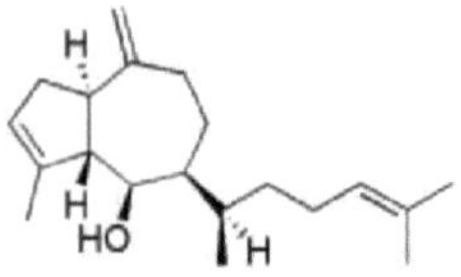

33

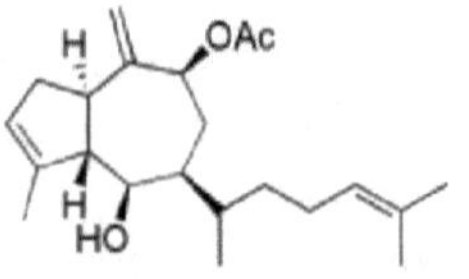

34

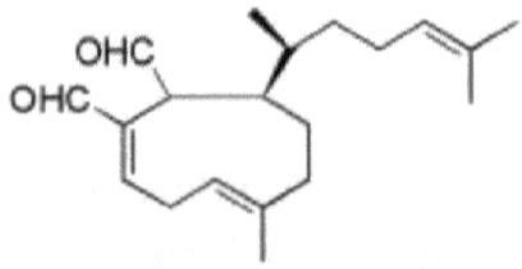

35

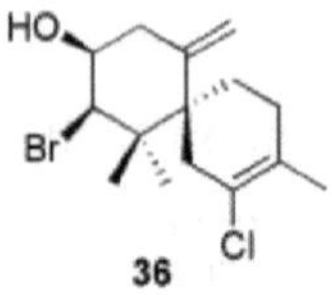

36

A combination of mechanisms may generally protect colonial ascidians from fouling. The Mediterranean didemnid *Polysyncraton lacazei* benefits from associational grazing by two arthropods, periodic sloughing of its surface cuticle every 4–15 months, and a diverse chemical defense that variously inhibits growth of bacteria, fungi, diatoms, and sea urchin larvae (Wahl and Banaigs 1991). Surface properties of the tunic were not responsible for antifouling defense. In a study of six colonial ascidians from the Mediterranean, palatability to predatory fish and crustaceans decreased with tissue energy content but was not associated with chemistry or toxicity (Tarjuelo et al. 2002). Physical mechanisms such as increased proportion of structural materials and lower energy content may thus contribute to antipredator defense in colonial ascidians. The degree to which physical properties of the tunic analogously affect patterns of fouling in solitary versus colonial ascidians remains to be elucidated.

The Atlantic species *Eudistoma olivaceum* produces a suite of over 20 alkaloids termed eudistomins (Rinehart et al. 1987). Early reports attributing the antifouling properties of this species to surface acidity or high levels of vanadium in the blood were subsequently refuted, and antilarval activity was traced to a pair of isomeric β-carboline alkoloids, eudistomins G (**29**) and H (**30**) (Stoecker 1978; Davis and Wright 1989; Davis et al. 1989). Both compounds inhibited settlement of *Bugula neritina* larvae at 2 μg per cm^2, 5% of the concentration of eudistomin H in the ascidian; however, these compounds were toxic to larvae of *Bugula* in bioassays (Davis and Wright 1990). A non-toxic inhibitor was detected but not isolated. In subsequent trials with larvae of two *Bugula* species, an ascidian and a barnacle, the eudistomin fraction inhibited settlement at 5% of the natural concentration. Compared to controls, panels treated with eudistomins were half as fouled by the ascidian *Diplosoma glandulosum* after a 7-h field deployment (Davis 1991). Eudistomins G and H were not concentrated in any part of the colony, did not deter fish feeding, and were only mildly antiviral compared to other eudistomins, implying a specific ecological role as antifoulants (Rinehart et al. 1987; Davis 1991).

The fouled solitary ascidian *Molgula occidentalis* was compared with two unfouled colonial species, *Amaroucium stellatum* and *Botryllus planus* (Bryan et al. 2003). Extracts of the unfouled species actually increased attachment of bacteria, yet the same extracts displayed antibacterial activity against five of six marine bacterial strains in standard disc assays. Extracts of *A. stellatum* were generally inhibitory to barnacle cyprid settlement in lab assays, whereas extracts of *M. occidentalis* were not. However, effects in laboratory and field trials were contradictory for extracts of *B. planus*. A chloroform extract inhibited barnacle settlement in the field, yet increased cyprid settlement in the lab; conversely, a methanol extract was inhibitory to barnacle settlement in the lab, but was non-effective in the field (Bryan et al. 2003). Chloroform extracts of all three species intensified fouling by bryozoans in the field, and no extract prevented fouling by polychaetes. These conflicting data point to the need for field-based evaluations of laboratory findings, as comparable results are not always obtained, particularly against a phylogenetically diverse assemblage of fouling larvae.

4.4 Bryozoans

Despite their calcified exterior, many bryozoan species maintain unfouled surfaces, yet antifouling defenses have rarely been investigated in this phylum. Walls et al. (1993) examined four Australian bryozoans for bacterial abundance, macrofouling, and the presence of antimicrobial chemistry. Bacteria and epibionts were less abundant on *Orthoscuticella ventricosa*, which produced antibiotic extracts, than on *Cellaria pilosa* and *Bugularia dissimilis*, which did not contain antimicrobial compounds. The pattern was obscured by *Amathia wilsoni*, which contained antibacterial secondary metabolites and was relatively unfouled, yet had the highest overall microbial abundance due to high-density patches. The possibility that select bacterial strains were attracted to this species was discussed but not investigated.

The congeners *Zoobotryon verticillatum* and *Z. pellucidum* contained 2,5,6-tribromo-1-methylgramine (**31**), an antibiotic and potent, non-toxic inhibitor of cyprid settlement (Sato and Fenical 1983; Kon-ya et al. 1994a). One of the most active antisettlement natural products, **31** was six times more inhibitory yet tenfold less toxic to cyprids than tributyltin oxide, and also blocked larval settlement of the mussel *Mytilus edulis* (Kon-ya et al. 1994a). Simpler indole derivatives also had strong antisettlement activity, repelling but not killing *Balanus* cyprids (Kon-ya et al. 1994b). Although many bryozoans are protected from fouling larvae by physical defenses such as pincer-like avicularia (Dyrynda 1986), encrusting species in particular deserve more attention for their potential chemical defenses and bacterial symbioses.

5 Investigating Larval Behavior to Understand and Combat Fouling

The structural complexity and biological activity of marine natural products, paired with the observation that few chemically rich species are fouled in the field, has led to the long-standing hypothesis that secondary metabolites defend invertebrates from fouling. Many natural products are toxic to the larvae of common fouling organisms, generally taken as support for the hypothesis that these compounds are antifouling in situ (Fusetani 2004). However, more knowledge is needed about larval behavior during settlement to interpret accurately the role of host chemistry in antifouling defense. Water-soluble, biological macromolecules can also inhibit fouling (Holmström and Kjelleberg 1993; James et al. 1996; Harder and Qian 2000), and despite low solubility in seawater, some non-polar metabolites diffuse from the surface of producing organisms. This raises fundamental questions about how larvae experience and process chemical deterrents or toxins in nature. Can larvae react to waterborne signals or toxins before contact with a potential basibiont, or are such compounds detected during surface exploration? Do antifouling defenses primarily repel larvae, or kill settled larvae and juveniles? To answer such questions, critical information is needed concerning:

(1) whether larvae perceive compounds in the water column, prior to surface contact
(2) the behavior of larvae around live, chemically rich surfaces
(3) the ecological significance of toxicity versus non-toxic deterrence

5.1 Waterborne Signals and Chemically Mediated Navigation

Historically, waterborne settlement cues were considered unimportant to the recruitment of invertebrate larvae in near-bottom flows, where turbulence would rapidly dilute dissolved compounds to levels below detection thresholds (Crisp 1974; Butman 1987; Pawlik 1992). Further, the slow swimming speeds and small size of larvae were thought to preclude chemotaxis, movement along a chemical gradient, from being effective under field conditions. However, larvae are now known to exhibit behavioral responses to waterborne signals that can enhance settlement success in flow. Dissolved compounds may also function as settlement deterrents, but most studies have hitherto focused on waterborne inducers rather than inhibitors.

The assumption that turbulence would prohibit larval chemotaxis misconstrues how chemically mediated navigation occurs in nature. Both air and water are inherently turbulent milieus, and the level of turbulence dictates how chemical signals propagate and disperse. A source of odorant produces plumes of molecules that are twisted and distorted by the ambient flow, creating filaments containing higher levels of the signal than the surrounding medium (Zimmer and Butman 2000). Many animals use a similar mechanism to navigate in turbulent odor plumes: perception of a chemical signal initiates orientation to a physical parameter, such as wind or current direction (Mafra-Neto and Cardé 1994; Nevitt et al. 1995). For example, upon sensing a pheromone plume, male moths immediately fly upwind; if they lose the plume, they cast back and forth until it is re-encountered (Vickers and Baker 1994; Mafra-Neto and Cardé 1994). Crabs track odor plumes from prey similarly, moving upstream against flow within a plume and tacking laterally to locate lost plumes (Weissburg and Zimmer-Faust 1994). Tracking behaviors evolved in turbulent settings where rheotaxis is a reliable way to trace a scent back to its source; such behavior is not true "chemotaxis", as a gradient is not necessary for navigation.

Analogously, dissolved cues can effectively mediate settlement by triggering downward displacement of larvae; when vertical distributions become bottom-skewed, more individuals are swept by turbulent eddies into contact with the bed (Turner et al. 1994; Tamburri et al. 1996). In still water, bivalve and gastropod larvae turned more frequently, decreased speed, and moved downwards when stimulated with waterborne cues (Tamburri et al. 1992; Krug and Zimmer 2000). In flume studies, larvae moved downwards and even swam upstream against flow after passing over a benthic source of settlement cue (Tamburri et al. 1996). Dissolved signals can thus mediate settlement by inducing larvae to orient and move towards the bed, using light or gravity for direction (Forward and Rittschof 1994; Eckman 1996; Tamburri et al. 1996). Indeed, controlled release of an inductive peptide from polyacrylamide gels increased recruitment of barnacle larvae in field assays (Browne and Zimmer 2001).

These studies provide a hypothesis for how waterborne antifouling agents might act. Since dissolved attractants cause larvae to turn more, slow down, and move towards the bed, settlement *inhibitors* should cause larvae to turn less, accelerate, and move upwards or away from a substratum. Such chemically mediated rejection will cause larvae to be swept by eddies or advected away from a repellant surface in the field. Larval behavior should thus be tested in simulated flows, yet surprisingly few studies of fouling behavior have been performed in flume tanks, where flow speed and turbulence can be manipulated as experimental variables. Cyprid settlement was studied in different flows (Mullineaux and Butman 1991), but larval behavior around chemically protected invertebrates has yet to be examined in a flume. The behavior of *Balanus improvisus* cyprids over smooth and textured surfaces was studied in still water and flume assays, testing behavioral rejection of substrata (Berntsson et al. 2004). In still water, cyprids moved faster, continuously and randomly over textured surfaces; however, conspecific extract resulted in slower, more twisted paths, indicating greater surface exploration. In a flume, the proportion of larvae remaining on textured surfaces was lower than on smooth surfaces after 5 min, as larvae that rejected a surface were swept away by the flow.

5.2
Measuring Production and Release of Chemical Deterrents in Situ

Field distributions of water-soluble molecules reflect rates of production and release, followed by transport of chemicals through advection and turbulent mixing (Zimmer and Butman 2000). To assess the ecological importance of dissolved cues, it is critical to define spatial and temporal scales at which waterborne molecules occur in field habitats (Jennings and Steinberg 1994). This is especially true when a compound is extracted from an intact organism. Molecules localized within the body of an invertebrate may function as antifeedants, acting after a predator punctures the tissue to deter subsequent attacks. However, metabolites present deep inside an organism may not be perceptible to larvae settling on the outer cell layer or cuticle.

Localizing compounds within invertebrate tissues or colonies is critical to the study of fouling defenses. Some sponge compounds are found in spherulous cells, which may be shed into the sponge's environment (Walker et al. 1985; Uriz et al. 1996). In other cases, metabolites are in the outer layers or distributed throughout a sponge (Kubanek et al. 2002). Metabolites produced by bacterial symbionts may remain inside the microbes (Unson and Faulkner 1993; Unson et al. 1994; Bewley et al. 1996). For instance, the bryostatins are localized inside a bacterial endosymbiont, which is housed inside the larvae, which are brooded within adult colonies (Davidson et al. 2001; Lopanik et al. 2004). Such compounds may not be biologically available for defense at the host's surface.

Despite indirect evidence that waterborne deterrents can inhibit fouling, bioactive compounds have rarely been quantified in seawater overlying invertebrate surfaces. Turbulent eddies dilute and disperse compounds as they are exuded from a point source, so even a continuously released metabolite will not readily accumulate in the water column. Measuring the effective in situ concentrations of putative deterrents is essential to test their ecological role. Few studies have even attempted to measure the concentration of potential antifoulants in seawater. Coll et al. (1982) presented qualitative evidence that non-polar terpenes were exuded by the soft corals *Sarcophyton crassocaule* and *Sinularia flexibilis*. Using a modified apparatus, Schulte et al. (1991) pumped seawater past an *S. flexibilis* colony and over Sep-pak cartridges, trapping sufficient flexibilide (**26**) to detect by TLC but not by NMR. Two methods failed to recover triterpene glycosides **7–8** from water around two Caribbean sponges, even though glycosides should be more soluble than a terpene like flexibilide (Kubanek et al. 2002). Clearly, this area is ripe for methodological innovation to quantify in situ rates of antifoulant release.

5.3
Behavior of Larvae around Chemically Defended Surfaces

To assess reliably whether a compound is an antifoulant, it is critical to study how larvae respond to natural concentrations under authentic conditions. In particular, there is strikingly little information on where rejection of a surface occurs. Larvae may detect a chemically defended organism prior to, or upon, initial contact; alternatively, metamorphosis could be blocked, or juveniles killed, by toxins leaching from a surface. Understanding such processes may determine, for instance, if antifouling paints cause post-settlement mortality whereas invertebrates trigger larval avoidance through waterborne or surface-bound deterrents. Given the importance of reduced exploratory behavior in settlement inhibition (Berntsson et al. 2004), we need to understand the mechanisms that govern site choice by fouling larvae.

Larvae are known to be extraordinarily sensitive to chemical cues during settlement, and as settling on a toxic surface is a fatal mistake, larvae have likely evolved mechanisms for detecting and rejecting poisoned surfaces (either before or upon contact). Evidence of such rejection is indirect, however, due to a lack of research into chemically triggered avoidance behavior. In the field, ascidian tadpole larvae made frequent contacts with the sponge *Mycale* sp. without settling, but almost never contacted the surface of *Crella incrustans* (Davis et al. 1991). As *C. incrustans* was not allelopathic to juvenile ascidians, it likely produces waterborne deterrents that act at a distance to inhibit larval contact. A promising innovation for use with smaller larvae is in situ video monitoring during substratum exploration, and subsequent analysis of behavior (Hills et al. 2000; Thomason et al. 2002). For instance, *Semibalanus balanoides* cyprids were video recorded in the field over three test substrata, and analysis of the video revealed differences in small-scale exploratory behavior corresponding to surface properties of each substratum (Thomason et al. 2002). This technique offers the promise of determining the scale at which larvae recognize and respond to physical and chemical properties of a surface, either attractive or repellent, under natural conditions of flow, conspecific abundance, etc.

The clearest evidence for larval rejection of defended surfaces comes from algae. The thinness of algal blades, compared to invertebrate surfaces, makes algae more amenable to behavioral observations in the laboratory. The brown alga *Dictyota menstrualis* was less fouled than other co-occurring algae, despite having more surface area, and contained diterpenes such as dictyol E (**32**) and pachydictyol A (**33**) (Schmitt et al. 1995). In lab assays, larvae of *Bugula neritina* contacted *D. menstrualis* at the same rate as preferred algae, but were 100-fold less likely to settle on *Dictyota*. Thus, larvae displayed no avoidance behavior prior to contacting the alga, where diterpenes are concentrated in a surface layer 1-cell thick. Dictyol E and dictyol B acetate (**34**) were highly toxic to larvae prior to settlement, when dissolved in seawater at a dose lower than the effective concentration on the algal surface; in contrast, pachydictyol A and dictyodiol (**35**) were not toxic to larvae, but slowed growth and deformed development of juveniles (Schmitt et al. 1995, 1998). None of the diterpenes was toxic to larvae of the bryozoan *Amathia convoluta*, but all inhibited metamorphosis and showed post-settlement toxicity at relevant concentrations; further, dictyol E and dictyol B acetate decreased settlement and juvenile development in the hydroid *Eudendrium carneum* (Schmitt et al. 1998). Thus, non-polar metabolites may not affect larvae prior to contact, but can diminish settlement and post-settlement survival of juvenile stages. Whether larvae spend less time exploring dictyol-treated surfaces or change behavior after contact with such a surface is unclear, but evidence suggests no avoidance behavior acts upstream of initial settlement.

In some algae, non-polar metabolites are concentrated at the surface (Dworjanyn et al. 1999). In others, high-molecular weight, biological macromolecules act as waterborne settlement inhibitors (Harder and Qian 2000), but chemical characterization of biomolecules has traditionally lagged far behind that of secondary metabolites. In a comparison of two fucoid algae, cyprid settlement was more inhibited by algal blades and phlorotannins of *Fucus vesiculoides* than *F. evanescens*, although *F. vesiculoides* was more heavily fouled in the field (Wikström and Pavia 2004). Post-settlement survival of barnacles was lower on *F. evanescens* due to higher rates of detachment, however; this highlights the need to delineate between pre-settlement deterrents and post-settlement mortality (due to absorption of toxins, inhibitors of metamorphosis, or physical factors).

5.4 Interpreting the Ecological Importance of Larval Toxicity in Laboratory Assays

Laboratory still-water assays have variously demonstrated that extracts of marine invertebrates can (1) inhibit settlement without otherwise impairing larvae, (2) cause developmental abnormalities in embryos, (3) exhibit sub-lethal toxicity to larvae, and (4) kill larvae or juveniles. However, the ecological relevance of such bioactivity is unclear. Marine invertebrates produce a dazzling variety of cytotoxic compounds, active against a wide range of organisms, embryos and cell lines (Faulkner 1984; Pettit et al. 1985). Given the number of metabolites that are cytotoxic or lethal to vertebrates at a sufficient dose, it is to be expected that many natural products are toxic to vulnerable embryonic or larval stages of other species. Killing larvae does not, however, indicate that a compound has an ecological function as an antifoulant, any more than an LD_{50} for mice indicates that a metabolite naturally functions to protect a sponge from rodents.

Broad-spectrum toxicity is a property of so many marine natural products that care should be exercised in extrapolating ecological function from still-water bioassays. Many natural products bind to specific intracellular protein targets that are highly conserved among eukaryotes, including kinases (Tosuji et al. 2003), telomerases (Warabi et al. 2003), proteases (Nakao et al. 1999, 2000; Fujita et al. 2001), and ion pumps (Sata et al. 1999). Such mechanisms of action are likely to lead to widespread bioactivity, and indeed, many compounds that inhibit settlement also kill larvae or have other cytotoxic properties. Practical considerations favor the use of standard, still-water bioassays using larvae of easily obtained fouling organisms. However, the widespread reliance on such methods and the lack of field trials has limited our understanding of how chemical defenses protect invertebrates against fouling in nature. Demonstrating that a compound inhibits settlement of fouling larvae in still water is suggestive of a natural role, and reason for further investigation, but does not *a priori* indicate an ecological function. Larvae are inherently mobile, and in the pervasive turbulence of benthic environments, will not remain localized near a surface unless they settle and attach to it. There is no natural situation analogous to larvae that are trapped in a small volume of water containing a chemical solution in the laboratory. Even if a given metabolite is present in the water column at a biologically relevant concentration, swimming larvae will at best be transiently exposed to pulses of a compound while searching for a settlement site.

Fouling is an ecological *process*, involving organisms from different kingdoms and interactions that develop over time. Laboratory assays using larvae of 1–2 test species do not assess whether a metabolite deters fouling in the extracted organism, and cannot be interpreted in an ecological context. I therefore recommend a shift in terminology. The term *antifouling* should be reserved for compounds tested in the field against a natural pool of microbes, algal spores and invertebrate larvae, or for which ecologically relevant data demonstrate a role against fouling in situ. Compounds that inhibit larval settlement in still-water bioassays should be termed *antisettlement*; this reflects their measured biological activity, leaving open the question of whether such activity may be effective against a diverse array of fouling organisms under field conditions. This is particularly relevant to studies showing reversible or non-toxic settlement inhibition by natural products. The term *antilarval* or *larvicidal* should be used for compounds that are toxic to larvae in bioassays, as these terms reflect the observed lethality to larval stages.

Compounds involved in antifouling defense may be broadly toxic, inhibiting larvae from completing metamorphosis via sub-lethal toxicity or killing settled larvae outright. The use of generally toxic compounds in antifouling defense may result from selection against the costs of synthesizing distinct defenses for different threats, leading to multiple ecological roles for a given metabolite (Kubanek et al. 2002). Alternatively, invertebrates may have chemical defenses shaped by natural selection specifically against fouling organisms. Such evolutionarily honed weapons would deter settlement or be *selectively* toxic to larvae, acting on sensory or internal biochemical pathways unique to larval stages, but would not affect embryos, cell lines, etc. A partnering of traditional, medically driven natural products research and ecologically motivated fouling studies may establish which of these hypothetical mechanisms predominates.

5.4.1 Behavioral Deterrence of Larvae Versus Metabolic Toxicity

Throughout the 1990's, increasing attention was paid to non-toxic inhibition of larval settlement, largely due to the commercial potential of environmentally safe alternatives to current paints (Fusetani et al. 1996; Fusetani 1997). Non-polar metabolites may repel larvae exploring a basibiont's surface; alternatively, polar metabolites liberated into overlying water may be detected by larval receptors and trigger avoidance behavior. Such antifouling chemicals should be classified as *deterrents*, repelling larvae via behavioral mechanisms before or during surface exploration, prior to the irreversible commitment to metamorphosis. Alternatively, waterborne or surface-adsorbed molecules may impair larvae through sub-lethal effects, or kill settled and metamorphosing stages outright. These would be *toxins*, negatively affecting larvae or killing those newly settled or metamorphosing on the organism's surface. Although the effects of sub-lethal toxicity may only be detectable over long periods (Ng and Keough 2003), some testable predictions emerge from this dichotomy that should be amenable to lab experimentation.

Non-toxic antifoulants may act on larval sensory systems, or inhibit early settlement responses without causing damage. Such a molecular defense is akin to bad-tasting but non-toxic antifeedants, like capsaicins from chili peppers or terpenes synthesized *de novo* by some dorid nudibranchs (Okuda et al. 1983; Krug et al. 1995). Non-toxic compounds may repel larvae through behavioral mechanisms, interfere with ligand binding of natural inducers, impair neurotransmission of inductive signals, or cause short-term ciliary arrest. Such metabolites do not likely act post-metamorphosis, and must therefore quickly deter larvae before or upon contact with the host. Marine natural products in this category often contain isocyano, furan, lactone, or bromine functional groups (Hirota et al. 1996; Clare et al. 1999; Sjögren et al. 2004).

One testable prediction is that non-toxic inhibitors may block settlement in the presence of some natural inductive cues, but not others. For example, the inhibitor phloroglucinol diminished only the gregarious component of barnacle settlement, a response to cues from other cyprids, and had no effect on isolated, individual larvae (Head et al. 2003). The selective inhibition of gregariousness among cyprids indicates a non-toxic mechanism of action for phloroglucinol, impairing recognition of conspecific cues without *a priori* impeding a larva's ability to metamorphose. Another prediction is that non-toxic deterrents should impede settlement in response to natural cues, but *not* in response to pharmacological inducers that act downstream of normal signal transduction pathways. This has not been well studied to date.

In contrast, toxic metabolites could have pronounced post-settlement effects. Although larvae may be adapted to recognize and avoid toxic coatings prior to attachment, surface-bound toxins may require longer contact times than non-toxic deterrents for their effects to manifest, and may act well after settlement (Schmitt et al. 1998). One prediction is that settled stages should be affected by toxins, but not by non-toxic deterrents. A second prediction is that toxins will block settlement in the presence of neuroactive compounds, interfering with the biochemical progression through metamorphosis that such agents artificially trigger. For instance, sub-lethal doses of 3 commercial pesticides (including DDT) blocked 100% of the settlement response in abalone veligers triggered by exogenous addition of the neurotransmitter GABA (Morse et al. 1979). Even at concentrations that do not cause direct mortality, toxins may interfere with the biochemical pathways necessary for normal metamorphosis to occur.

Metals in antifouling paints may not be detectable by larval chemoreceptors, and likely act after larvae have contacted, and potentially attached to, treated surfaces. Indeed, $CuSO_4$-treated surfaces did not deter initial exploration or byssal thread attachment by mussel larvae, but eventually affected larval behavior after 2 hr of exposure, and killed 10% of larvae after 24 h (da Gama et al. 2003). In contrast, extracts of the unfouled alga *Laurencia obtusa* containing the terpene elatol (**36**) were immediately repellant to larvae and deterred attachment, and showed no toxicity for 24 h. Elatol is toxic to barnacle nauplii, bryozoan larvae, fish and insects (Hay et al. 1987; de Nys et al. 1996; König and Wright 1997), yet triggered behavioral rejection of surfaces without observed toxicity in mussel pediveligers. These results suggest larvae have adapted to perceive and avoid algal toxins, which therefore act as settlement deterrents, whereas metals are not detected and therefore act via delayed toxicity.

5.5 An Alternative Bioassay Design to Avoid Artifacts and Concentration Effects

A problem with laboratory bioassays is their reliance on static conditions; virtually all are performed in small volumes of still water. Such designs hinder the development of ecologically meaningful datasets. In still-water assays, larvae experience higher concentrations of toxins than they would encounter in nature, whether the compounds are pre-dissolved into the assay water or first dried onto a surface and allowed to diffuse into seawater. High levels of antifouling metabolites have never been measured in seawater around a chemically defended invertebrate in the field, where turbulence and advection will dilute and remove compounds exuded from live surfaces. Testing defined concentrations does not even permit comparison of EC_{50} values between studies when different larval densities were used in assays (Head et al. 2003). A more subtle problem may be the lack of hydrodynamics as a stimulus to larvae settling in still water. Flow and chemical cues can interactively determine settlement success (Pawlik and Butman 1993). As most fouling organisms are suspension feeders, their larvae may respond to flow as a cue during habitat choice; assays in static water may fail to trigger natural behaviors. This is an intrinsic limitation on the usefulness of larval bioassays from an ecological perspective, as it is difficult to extrapolate from unnatural assay conditions to the field.

As an alternative to commonly used assay dishes, plastic enclosures could be fashioned with mesh sides of 100-μm pore size to retain larvae, and a hollow bottom that can be filled with a polymeric gel. Compounds are embedded in a matrix of Phytogel or acrylamide, from which they diffuse at a controlled and calculable rate from the bottom of the container (Henrikson and Pawlik 1995; Browne and Zimmer 2001). The concentration in the gel can be set to match that in an organism, or a rate of release measured in the field. The enclosure is then immersed in seawater, larvae added, and the lid sealed. The whole container is submerged in a large volume of seawater in a flow-through or filtered, recirculating aquarium. Larvae will be exposed to metabolites that are diffusing from the gel, but are free to move away from the gel surface, limiting their exposure to toxins; compounds are continuously diluted and removed by water passing through the mesh sides. This more accurately mimics field conditions, where larvae are not trapped in water with high concentrations of dissolved organic compounds.

To test whether contact with a treated surface is necessary, the gel surface can be covered with crushed shell to which larvae can attach; larvae and settlers will still be exposed to diffusing compounds, but not through direct contact with the surface. This will determine whether a chemical of interest acts as a waterborne or surface-adsorbed agent. Recently settled larvae and juveniles could easily be introduced into the containers, to measure post-settlement effects of metabolites diffusing from the gel. Such assays will be more ecologically meaningful and should provide new insight into how chemical antifouling defenses operate in nature.

Additional factors should be taken into consideration for bioassay design. The importance of density-dependent effects is clear for larvae of species that exhibit gregarious settlement and kin recognition, such as barnacles and bryozoans (Keough 1989). Gregariousness can affect assays at concentrations as low as 5 barnacle cyprids per dish, and shows unexpected interactions with settlement inducers and inhibitors (Head et al. 2003). Larvae from multiple batches should be used, to examine variation between batches as a factor (Raimondi and Keough 1990; Willemsen et al. 1998). Variation is an important issue in larval biology but has been less considered in fouling research (Holm 1990). Cyprid settlement response can vary over the course of a recruitment season (Jarrett 1997), and other species have inherently dimorphic settlement requirements (Toonen and Pawlik 2001); the impact of such behavioral variance on response to chemically protected surfaces remains unknown. Similarly, whether fouling larvae are locally adapted to the chemistry of common basibiont species remains to be tested; indeed, potential "resistance" to antifouling chemical defense has gone largely unexplored.

Methods using computer-assisted motion analysis to quantify behavior of larvae have been developed, but hitherto have been primarily used to study responses to settlement cues rather than inhibitors (Tamburri et al. 1992, 1996; Krug and Zimmer 2000). However, motion analysis has been used to quantify the behavior of algal spores and diatoms for antifoulant screening programs (Wigglesworth-Cooksey and Cooksey 1996; Iken et al. 2001, 2003); such approaches could be widely applied to study the behavior of fouling larvae during settlement.

In summary, the design of antilarval and antisettlement assays should pay careful attention to the realism of conditions to which larvae are exposed. Larval behavior should be observed and quantified where possible, and the reversibility of inhibition or lethality of treatments assessed. Ideally, assays should expose larvae only to concentrations likely to be encountered in the field, either by reproducing concentrations measured in situ, or by allowing compounds to diffuse into an excess of seawater while larvae are confined near a treated surface. Flume and field experiments may provide missing information on larval behavior around chemically enriched surfaces in realistic flow.

6 The Importance of Alternative Hypothesis Testing: Mechanical and Physical Defense

Science is founded on the testing and rejection of unsupported hypotheses. It is common in the literature for unusual chemistry to be found in an unfouled organism, and if the chemistry is active in antilarval bioassays, a single conclusion is asserted: the chemistry serves an antifouling purpose in nature. This is too often the end of the story, rather than a starting point; thorough testing of many hypotheses is needed to find the best explanation for how an organism maintains a clean surface. Examples of robust testing of physical, mechanical, behavioral and chemical defense hypotheses can be found in Wahl and Banaigs (1991), Becker and Wahl (1996), and Wahl et al. (1998). In these studies, the authors rejected all but one or two possible antifouling defenses for each species examined, sometimes finding that a combination of different strategies was employed (e.g., sloughing and chemical defense in the ascidian *Polysyncraton lacazei*; Wahl and Banaigs 1991).

A wide range of mechanisms can reduce fouling in invertebrates, yet are frequently overlooked in favor of chemical defenses. Given that mechanical, behavioral and chemical defenses may operate in concert, it is critical to test alternatives to a strictly chemical model of antifouling. Mobile organisms can protect their surfaces through a combination of mechanisms. For example, studies of seven tropical and one temperate crab species found that carapaces were kept unfouled by a combination of behavioral and mechanical strategies, including burrowing to scrape surfaces clean, nocturnal activity patterns that inhibit algal growth, and time spent in air (Becker and Wahl 1996; Wahl et al. 1998). Mutual surface grazing acts as a density-dependent defense in some snail populations (Wahl and Sonnichsen 1992).

Sessile species have fewer non-chemical options, but physical defenses should be considered as alternative hypotheses in any investigation of antifouling (Dyrynda 1986). Microtopography may impede attachment of spores and larvae (Wahl et al. 1998; Scardino et al. 2003; Bers and Wahl 2004). Mechanical defenses include features such as spicules, or miniature pinchers such as echinoderm pedicellaria or bryozoan avicularia (Dyrynda 1986); however, due to size constraints, these structures are only effective against larger organisms such as larvae. Movement of cytoplasm or cells from fouled regions to areas of new growth protects algae (Littler and Littler 1999), and could possibly function in sponges and colonial organisms.

Regular sloughing of outer layers protects certain algae (Filion-Myklebust and Norton 1981; Sieburth and Tootle 1981; Russell and Veltkamp 1984) and sponges (Barthel and Wolfrath 1989) from overgrowth. The study by Barthel and Wolfrath is noteworthy: it is only 3 pages long, yet one of the most cited papers in the fouling literature. More studies on sloughing are clearly needed to determine the general importance of this mechanism. Periodic molting in ecdysozoans or shedding of the tunic in ascidians protects against later stages of fouling such as colonization by invertebrate larvae, but is too infrequent to deter unicellular epibionts (Wahl and Banaigs 1991; Becker and Wahl 1996). Regular shedding of mucus layers may be a significant antifouling mechanism in cnidarians (Johannes 1967; Ducklow and Mitchell 1979). Removal of fouling larvae by suspension feeding has been suggested (Cowden et al. 1984), and bivalve aggregations may reduce fouling pressure by filtering larvae out of the water column, an effective if indirect defense (Tamburri and Zimmer-Faust 1996; Wahl et al. 1998).

The tight degree of association occasionally observed for specific epibiont-basibiont pairs remains relatively unexplored (Davis et al. 1996). For instance, Davis and White (1994) found distinctive associations between an ascidian and an encrusting sponge, a sponge and a zoanthid, and a bryozoan and an anthozoan. Little has been done to investigate the mechanistic basis for species-specific attraction of epibiont larvae in these presumptive mutualisms, which if co-evolved may not incur the typical costs of fouling to the basibiont. One clear advantage would thus be defense against the broader array of potential fouling organisms, whose larvae may be spatially excluded, chemically repelled, or consumed by the host-associated epibionts.

7 Conclusions

Marine invertebrates utilize an extraordinary range of antifouling strategies, and any given species likely derives some benefit from multiple defense mechanisms. Chemistry can either kill or repel bacteria, or slow the growth of diatoms, thereby removing positive settlement cues that would otherwise encourage attachment of larvae. More coevolved defenses occur where invertebrates selectively attract host-specific microbes to their surfaces, which then displace or inhibit competing bacteria, or chemically deter larval settlement. Basibionts can also directly defend against fouling organisms, either via non-toxic deterrents that induce avoidance behavior in larvae, or through toxins that repel exploring larvae or kill recently attached settlers.

Improved bioassays that allow larvae more realistic freedom of movement and expose them to authentic doses of natural chemicals may shed greater light on how the chemical defenses of marine organisms operate in the field. Quantitative observations of larval behavior should be incorporated into antifouling screening, to provide more sophisticated insight into how larvae act around chemically imbued surfaces, particularly in moving water. Strikingly, it is still unclear whether most natural chemical defenses work by deterring larvae prior to contact, upon initial contact, or through post-settlement toxicity; more comparative studies will be needed to uncover the generality of a given mechanism, and which strategies are effective against particular fouling threats.

Evolutionarily, it would be interesting to learn whether chemical defenses are co-adapted to the regional fouling organisms that threaten a particular organism. However, global fouling communities have become highly homogenized, as human activity moves fouling larvae in ballast water and adult communities on hulls around the world (Minchin and Gollasch 2003). This may diminish the import of regional specialization in chemical defense against epibiosis before we have a chance to study it, much as invasive species generally threaten biodiversity and overwhelm the fragile adaptations of endemics. The ecological complexity of the interactions and the opportunities for mutualisms and evolutionary arms races between adult invertebrates, microbes, protists and larvae make the study of biofouling an intriguing arena for biologists and chemists alike. Hopefully we may learn from the marvelously adapted invertebrates of the benthos and devise environmentally sound antifouling practices, lest we do further harm to coastal ecosystems in our efforts to ward off the devilishly persistent propagules of fouling organisms.

Acknowledgements. I wish to thank Kenneth Boyd and Ute Hentschel for providing reprints, Demian Willette and Ryan Ellingson for assistance in manuscript preparation, and Alison McCurdy for access to software. Special thanks are due to Demian Willette for contributing artwork to Fig. 1. Much stimulating discussion came from participants on the chemical ecology UNOLS cruise on the R/V Seward Johnson (NSF award OCE-0095724 to JR Pawlik). The writing of this chapter was supported by NSF awards to PJK (OCE 02-42272 and HRD 03-17772).

References

Aceret TL, Sammarco PW, Coll JC (1995) Effects of diterpenes derived from the soft coral *Sinularia flexibilis* on the eggs, sperm and embryos of the scleractinian corals *Montipora digitata* and *Acropora tenuis*. Mar Biol 122:317–323

Allison DG (2003) The biofilm matrix. Biofouling 19:139–150

Amade P, Chevolot L (1982) Antimicrobial activities of marine sponges from French Polynesia and Brittany. Mar Biol 70:223–228

Armstrong E, Boyd KG, Pisacane A, Peppiatt CJ, Burgess JG (2000) Marine microbial natural products in antifouling coatings. Biofouling 16:215–224

Armstrong E, Yan LM, Boyd KG, Wright PC, Burgess JG (2001) The symbiotic role of marine microbes on living surfaces. Hydrobiologia 461:37–40

Avelinmary SX, Vitalinamary SX, Rittschof D, Nagabhushanam R (1993) Bacterial-barnacle interaction - potential of using juncellins and antibiotics to alter structure of bacterial communities. J Chem Ecol 19:2155–2167

Bahamondes-Rojas I, Dherbomez M (1990) Purification partielle des substances glycoconjuguees capables d'induire la metamorphose des larves competentes d'*Eubranchus doriae* (Trinchese, 1879), mollusques nudibranche. J Exp Mar Biol Ecol 144:17–27

Baier R (1981) Early events of micro-biofouling of all heat transfer equipment. In: Somerscules E, Knudsen J (eds) Fouling of heat transfer equipment. Hemisphere Publ, Washington, DC, pp 293–304

Baldwin IT, Preston CA (1999) The eco-physiological complexity of plant responses to insect herbivores. Planta 208:137–145

Bandurraga MM, Fenical W (1985) Isolation of the muricins: evidence of a chemical adaptation against fouling in the marine octocoral *Muricea fruticosa* (Gorgonacea). Tetrahedron 41:1057–1065
Barkai A, McQuaid C (1988) Predator–prey role reversal in a marine benthic ecosystem. Science 242:62–64
Barthel D, Wolfrath B (1989) Tissue sloughing in the sponge *Halichondria panicea*: a fouling organism prevents being fouled. Oecologia 78:357–360
Bauer WD, Robinson JB (2002) Disruption of bacterial quorum sensing by other organisms. Curr Opin Biotechnol 13:234–237
Becerro M, Uriz M, Turon X (1997) Chemically mediated interactions in benthic organisms: the chemical ecology of *Crambe crambe* (Porifera, Poecilosclerida). Hydrobiologia 356:77–89
Becker K, Wahl M (1996) Behaviour patterns as natural antifouling mechanisms of tropical marine crabs. J Exp Mar Biol Ecol 203:245–258
Bergquist RP, Bedford JJ (1978) The incidence of antibacterial activity in marine Demospongiae: systematic and geographic considerations. Mar Biol 46:215–221
Berking S (1987) Homarine (*N*-methylpicolinic acid) and trigonelline (*N*- methyl-nicotinic acid) appear to be involved in pattern control in a marine hydroid. Development 99:211–220
Berntsson KM, Jonsson PR, Larsson AI, Holdt S (2004) Rejection of unsuitable substrata as a potential driver of aggregated settlement in the barnacle *Balanus improvisus*. Mar Ecol Prog Ser 275:199–210
Bers AV, Wahl M (2004) The influence of natural surface microtopographies on fouling. Biofouling 20:43–51
Bewley CA, Holland ND, Faulkner DJ (1996) Two classes of metabolites from *Theonella swinhoei* are localized in distinct populations of bacterial symbionts. Experientia 52:716–722
Bloom SA (1975) The motile escape response of a sessile prey: a sponge-scallop mutualism. J Exp Mar Biol Ecol 17:311–321
Bobzin S, Faulkner D (1992) Chemistry and chemical ecology of the Bahamian sponge *Aplysilla glacialis*. J Chem Ecol 18:309–332
Boyd KG, Adams DR, Burgess JG (1999) Antibacterial and repellent activities of marine bacteria associated with algal surfaces. Biofouling 14:227–236
Brancato M, Woollacott R (1982) Effect of microbial films on settlement of bryozoan larvae (*Bugula simplex*, *B. stolonifera* and *B. turrita*). Mar Biol 71:51–56
Browne KA, Zimmer RK (2001) Controlled field release of a waterborne chemical signal stimulates planktonic larvae to settle. Biol Bull 200:87–91
Bryan PJ, Rittschof D, Qian PY (1997) Settlement inhibition of bryozoan larvae by bacterial films and aqueous leachates. Bull Mar Sci 61:849–857
Bryan PJ, McClintock JB, Slattery M, Rittschof DP (2003) A comparative study of the non-acidic chemically mediated antifoulant properties of three sympatric species of ascidians associated with seagrass habitats. Biofouling 19:235–245
Bulthuis DA, Woelkerling WJ (1983) Biomass accumulation and shading effects of epiphytes on leaves of seagrass, *Heterozostera tasmanica*, in Victoria, Australia. Aquat Bot 16:137–148
Burgess JG, Jordan EM, Bregu M, Mearns-Spragg, Boyd KG (1999) Microbial antagonism: a neglected avenue of natural products research. J Biotechnol 70:27–32
Burkholder PR, Rutzler K (1969) Antimicrobial activity of some marine sponges. Nature 222:983–984
Butman C (1987) Larval settlement of soft-sediment invertebrates: the spatial scales of pattern explained by active habitat selection and the emerging role of hydrodynamical processes. Oceanogr Mar Biol Annu Rev25:113–165
Carr WES, Netherton JC, Gleeson RA, Derby CD (1996) Stimulants of feeding behavior in fish: analyses of tissues of diverse marine organisms. Biol Bull 190:149–160

Clare AS (1996) Marine natural product antifoulants: status and potential. Biofouling 9: 211–229
Clare AS, Matsumura K (2000) Nature and perception of barnacle settlement pheromones. Biofouling 15:57–71
Clare AS, Rittschof D, Gerhart DJ, Hooper IR, Bonaventura J (1999) Antisettlement and narcotic action of analogues of diterpene marine natural product antifoulants from octocorals. Mar Biotechnol 1:427–436
Coll JC, Bowden BF, Tapiolas DM, Dunlap WC (1982) In situ isolation of allelochemicals released from soft corals (Coelenterata: Octocorallia): a totally submersible sampling apparatus. J Exp Mar Biol Ecol 60:293–299
Coll J, Price I, Konig G, Bowden B (1987) Algal overgrowth of alcyonacean soft corals. Mar Biol 96:129–135
Compere C, Bellon-Fortaine M-N, Bertrand P, Costa D, Marcus P, Poleunis C, Pradier C-M, Rondot B, Walls MG (2001) Kinetics of conditioning layer formation on stainless steel immersed in seawater. Biofouling 17:129–145
Cowden C, Young CM, Chia F-S (1984) Differential predation on marine invertebrate larvae by two benthic predators. Mar Ecol Prog Ser 14:145–149
Crisp DJ (1974) Factors influencing the settlement of marine invertebrate larvae. In: Grant PT, Machie AM (eds) Chemoreception in marine organisms. Academic Press, New York, pp 177–265
Crisp DJ, Meadows PS (1963) Adsorbed layers: the stimulus to settlement in barnacles. Proc R Soc Lond B 158:364–387
Cuba TR, Blake NJ (1983) The initial development of a marine fouling assemblage on a natural substrate in a subtropical estuary. Bot Mar 26:259–264
Da Gama BAP, Pereira RC, Soares AR, Teixeira VL, Yoneshigue-Valentin Y (2003) Is the mussel test a good indicator of antifouling activity? A comparison between laboratory and field assays. Biofouling 19:161–169
Daume S, Brand-Gardner S, Woelkerling W (1999) Preferential settlement of abalone larvae: diatom films versus non-geniculate corraline red algae. Aquaculture 174: 243–254
Davidson SK, Haygood MG (1999) Identification of sibling species of the bryozoan *Bugula neritina* that produce different anticancer bryostatins and harbor distinct strains of the bacterial symbiont "*Candidatus* Endobugula sertula". Biol Bull 196:273–280
Davidson SK, Allen SW, Lim GE, Anderson CM, Haygood MG (2001) Evidence for the biosynthesis of bryostatins by the bacterial symbiont "*Candidatus* Endobugula sertula" of the bryozoan *Bugula neritina*. Appl Environ Microbiol 67:4531–4537
Davies DG, Parsek MR, Pearson JP, Iglewski BH, Costerton JW, Greenberg EP (1998) The involvement of cell-to-cell signalling in the development of a bacterial biofilm. Science 280:295–298
Davis AR (1991) Alkaloids and ascidian chemical defense: evidence for the ecological role of natural products from *Eudistoma olivaceum*. Mar Biol 111:375–379
Davis AR, Wright AE (1989) Interspecific differences in fouling of two congeneric ascidians (*Eudistoma olivaceum* and *Eudistoma capsulatum*): is surface-acidity an effective defense. Mar Biol 102:491–497
Davis AR, Wright AE (1990) Inhibition of larval settlement by natural products from the ascidian, *Eudistoma olivaceum* (Van Name). J Chem Ecol 16:1349–1357
Davis AR, White GA (1994) Epibiosis in a guild of sessile subtidal invertebrates in south-eastern Australia: a quantitative survey. J Exp Mar Biol Ecol 177:1–14
Davis AR, Targett NM, McConnell OJ, Young CM (1989) Epibiosis of marine algae and benthic invertebrates: natural products chemistry and other mechanisms inhibiting settlement and overgrowth. In: Scheuer PJ (ed) Bioorganic marine chemistry. Springer, Berlin Heidelberg New York, pp 85–114
Davis AR, Butler AJ, Vanaltena I (1991) Settlement behavior of ascidian larvae: preliminary evidence for inhibition by sponge allelochemicals. Mar Ecol Prog Ser 72:117–123

Davis AR, Ayre DJ, Billingham MR, Styan CA, White GA (1996) The encrusting sponge *Halisarca laxus*: population genetics and association with the ascidian *Pyura spinifera*. Mar Biol 126:27–33

De Moraes CM, Lewis WJ, Pare PW, Alborn HT, Tumlinson JH (1998) Herbivore-infested plants selectively attract parasitoids. Nature 393:570–573

De Nys R, Leya T, Maximilien R, Afsar A, Nair PSR, Steinberg PD (1996) The need for standardised broad scale bioassay testing: a case study using the red alga *Laurencia rigida*. Biofouling 10:213–224

Dixon J, Schroeter SC, Kastendiek J (1981) Effects of encrusting bryozoan, *Membranipora membranacea*, on the loss of blades and fronds by the giant kelp, *Macrocystis pyrifera* (Laminariales). J Phycol 17:341–345

Dobretsov SV, Qian PY (2002) Effect of bacteria associated with the green alga *Ulva reticulata* on marine micro- and macrofouling. Biofouling 18:217–228

Dobretsov S, Qian PY (2004) The role of epibotic bacteria from the surface of the soft coral *Dendronephthya* sp in the inhibition of larval settlement. J Exp Mar Biol Ecol 299: 35–50

Dobretsov S, Dahms HU, Qian PY (2004) Antilarval and antimicrobial activity of waterborne metabolites of the sponge *Callyspongia* (*Euplacella*) *pulvinata*: evidence of allelopathy. Mar Ecol Prog Ser 271:133–146

Ducklow HW, Mitchell R (1979) Bacterial populations and adaptations in the mucus layers on living corals. Limnol Oceanogr 24:715–725

Dworjanyn SA, De Nys R, Steinberg PD (1999) Localisation and surface quantification of secondary metabolites in the red alga *Delisea pulchra*. Mar Biol 133:727–736

Dyrynda P (1986) Defensive strategies of modular organisms. Philos Trans R Soc Lond B 313:227–243

Eckman JE (1996) Closing the larval loop: linking larval ecology to the population dynamics of marine benthic invertebrates. J Exp Mar Biol Ecol 200:207–237

Egan S, Thomas T, Holmström C, Kjelleberg S (2000) Phylogenetic relationship and antifouling activity of bacterial epiphytes from the marine alga *Ulva lactuca*. Environ Microbiol 2:343–347

Enderlein P, Moorthi S, Rohrscheidt H, Wahl M (2003) Optimal foraging versus shared doom effects: interactive influence of mussel size and epibiosis on predator preference. J Exp Mar Biol Ecol 292:231–242

Engel S, Jensen PR, Fenical W (2002) Chemical ecology of marine microbial defense. J Chem Ecol 28:1971–1985

Faimali M, Sepčić K, Turk T, Geraci S (2003) Non-toxic antifouling activity of polymeric 3-alkylpyridinium salts from the Mediterranean sponge *Reniera sarai* (Pulitzer- Finali). Biofouling 19:47–56

Faulkner DJ (1984) Marine natural products – metabolites of marine invertebrates. Nat Prod Rep 1:551–598

Fearon RJ, Cameron AM (1997) Preliminary evidence supporting the ability of hermatypic corals to affect adversely larvae and early settlement stages of hard coral competitors. J Chem Ecol 23:1769–1780

Feifarek BP (1987) Spines and epibionts as antipredator defenses in the thorny oyster *Spondylus americanus* Hermann. J Exp Mar Biol Ecol 105:39–56

Fenical W (1993) Chemical studies of marine bacteria – developing a new resource. Chem Rev 93:1673–1683

Fieseler L, Horn M, Wagner M, Hentschel U (2004) Discovery of the novel candidate phylum "*Poribacteria*" in marine sponges. Appl Environ Microbiol 70:3724–3732

Filion-Myklebust C, Norton T (1981) Epidermis shedding in the brown seaweed *Ascophyllum nodosum* (L.) Le Jolis and its ecological significance. Mar Biol Lett 2:45–51

Fishlyn DA, Phillips DW (1980) Chemical camouflaging and behavioral defense against a predatory seastar by three species of gastropods from the surf grass *Phyllospadix* community. Biol Bull 158:34–48

Fletcher M, McEldowney S (1984) Microbial attachment to non-biological surfaces. In: Klug M, Reddy C (eds) Current perspectives in microbial ecology. Proc 3rd Int Symp Microb Ecol, 1983, pp 124–129
Forester AJ (1979) The association between the sponge *Halichondria panicea* (Pallas) and scallop *Chlamys varia* (L.): a commensal-protective mutualism. J Exp Mar Biol Ecol 36:1–10
Forward RB, Rittschof D (1994) Photoresponses of crab megalopae in offshore and estuarine waters: implications for transport. J Exp Mar Biol Ecol 182:183–192
Fujita M, Nakao Y, Matsunaga S, Seiki M, Itoh Y, van Soest RWM, Fusetani N (2001) Ancorinosides B-D, inhibitors of membrane type 1 matrix metalloproteinase (MT1-MMP), from the marine sponge *Penares sollasi* Thiele. Tetrahedron 57:1229–1234
Fusetani N (1997) Marine natural products influencing larval settlement and metamorphosis of benthic invertebrates. Curr Org Chem 1:127–152
Fusetani N (2004) Biofouling and antifouling. Nat Prod Rep 21:94–104
Fusetani N, Hirota H, Okino T, Tomono Y, Yoshimura E (1996) Antifouling activity of isocyanoterpenoids and related compounds isolated from a marine sponge and nudibranchs. J Nat Toxins 5:249–259
Geesey GG (2001) Bacterial behavior at surfaces. Curr Opin Microbiol 4:296–300
Gerhart D, Rittschof D, Mayo S (1988) Chemical ecology and the search for marine antifoulants. Studies of a predator-prey symbiosis. J Chem Ecol 14:1905–1917
Gil-Turnes MS, Fenical W (1992) Embryos of *Homarus americanus* are protected by epibiotic bacteria. Bioll Bull 182:105–108
Gil-Turnes MS, Hay ME, Fenical W (1989) Symbiotic marine bacteria chemically defend crustacean embryos from a pathogenic fungus. Science 246:116–118
Hadfield MG, Unabia CRC, Smith CM, Michael TM (1994) Settlement preferences of the ubiquitous fouler *Hydroides elegans*. In: Thompson MF, Nagabhushanam R, Sarojini R, Fingerman M (eds) Recent developments in biofouling control. Oxford and IBH, New Delhi, pp 65–74
Hall-Stoodley L, Stoodley P (2002) Developmental regulation of microbial biofilms. Curr Opin Biotechnol 13:228–233
Harder T, Qian PY (2000) Waterborne compounds from the green seaweed *Ulva reticulata* as inhibitive cues for larval attachment and metamorphosis in the polychaete *Hydroides elegans*. Biofouling 16:205–214
Harder TN, Thiyagarajan V, Qian PY (2001) Effect of cyprid age on the settlement of *Balanus amphitrite* Darwin in response to natural biofilms. Biofouling 17:211–219
Harder T, Lau SCK, Dahms HU, Qian PY (2002a) Isolation of bacterial metabolites as natural inducers for larval settlement in the marine polychaete *Hydroides elegans* (Haswell). J Chem Ecol 28:2029–2043
Harder T, Lam C, Qian PY (2002b) Induction of larval settlement in the polychaete *Hydroides elegans* by marine biofilms: an investigation of monospecific diatom films as settlement cues. Mar Ecol Prog Ser 229:105–112
Harder T, Lau SCK, Dobretsov S, Fang TK, Qian PY (2003) A distinctive epibiotic bacterial community on the soft coral *Dendronephthya* sp. and antibacterial activity of coral tissue extracts suggest a chemical mechanism against bacterial epibiosis. FEMS Microbiol Ecol 43:337–347
Hay M (1996) Marine chemical ecology: what's known and what's next? J Exp Mar Biol Ecol 200:103–134
Hay M, Duffy J, Fenical W (1987) Seaweed chemical defense against diverse coral reef herbivores. Ecology 68:1581–1591
Head RM, Overbeke K, Klijnstra J, Biersteker R, Thomason JC (2003) The effect of gregariousness in cyprid settlement assays. Biofouling 19:269–278
Henrikson AA, Pawlik JR (1995) A new antifouling assay method: results from field experiments using extracts of four marine organisms. J Exp Mar Biol Ecol 194:157–165

Hentschel U, Hopke J, Horn M, Friedrich AB, Wagner M, Hacker J, Moore BS (2002) Molecular evidence for a uniform microbial community in sponges from different oceans. Appl Environ Microbiol 68:4431–4440
Hills JM, Thomason JC, Davis H, Kohler J, Millett E (2000) Exploratory behaviour of barnacle larvae in field conditions. Biofouling 16:171–179
Hirota H, Tomono Y, Fusetani N (1996) Terpenoids with antifouling activity against barnacle larvae from the marine sponge *Acanthella cavernosa*. Tetrahedron 52:2359–2368
Hirota H, Okino T, Yoshimura E, Fusetani N (1998) Five new antifouling sesquiterpenes from two marine sponges of the genus *Axinyssa* and the nudibranch *Phyllidia pustulosa*. Tetrahedron 54:13971–13980
Holm ER (1990) Attachment behavior in the barnacle *Balanus amphitrite amphitrite* (Darwin): genetic and environmental effects. J Exp Mar Biol Ecol 135:85–98
Holmström C, Kjelleberg S (1993) Development of a tissue cell culture bioassay for identifying mechanisms of inhibition of settlement of barnacle and tunicate larvae by surface-colonizing marine bacteria. Biofouling 7:329–337
Holmström C, Kjelleberg S (1994) The effect of external biological factors on settlement of marine invertebrates and new antifouling technology. Biofouling 8:147–160
Holmström C, Kjelleberg S (1999) Marine *Pseudoalteromonas* species are associated with higher organisms and produce biologically active extracellular agents. FEMS Microbiol Ecol 30:285–293
Holmström C, Kjelleberg S (2000) Bacterial interaction with marine fouling organisms. In: Evans LV (ed) Biofilms: recent advances in their study and control. Harwood Academic Publ, Amsterdam, pp 101–115
Holmström C, Rittschof D, Kjelleberg S (1992) Inhibition of settlement by larvae of *Balanus amphitrite* and *Ciona intestinalis* by a surface-colonizing marine bacterium. Appl Environ Microbiol 58:2111–2115
Holmström C, James S, Egan S, Kjelleberg S (1996) Inhibition of common fouling organisms by marine bacterial isolates with special reference to the role of pigmented bacteria. Biofouling 10:251–259
Holmström C, Egan S, Franks A, McCloy S, Kjelleberg S (2002) Antifouling activities expressed by marine surface associated *Pseudoalteromonas* species. FEMS Microbiol Ecol 41: 47–58
Huang SY, Hadfield MG (2003) Composition and density of bacterial biofilms determine larval settlement of the polychaete *Hydroides elegans*. Mar Ecol Prog Ser 260: 161–172
Iken K, Amsler CD, Greer SR, McClintock JB (2001) Qualitative and quantitative studies of the swimming behaviour of *Hincksia irregularis* (Phaeophyceae) spores: ecological implications and parameters for quantitative swimming assays. Phycologia 40:359–366
Iken K, Greer SP, Amsler CD, McClintock JB (2003) A new antifouling bioassay monitoring brown algal spore swimming behaviour in the presence of echinoderm extracts. Biofouling 19:327–334
Ito S, Kitamura H (1997) Induction of larval metamorphosis in the sea cucumber *Stichopus japonicus* by periphitic diatoms. Hydrobiologia 358:281–284
James S, Holmström C, Kjelleberg S (1996) Purification and characterization of a novel antibacterial protein from the marine bacterium D2. Appl Environ Microbiol 62: 2783–2788
Jarrett JN (1997) Temporal variation in substratum specificity of *Semibalanus balanoides* (Linnaeus) cyprids. J Exp Mar Biol Ecol 211:103–114
Jennings JG, Steinberg PD (1994) In situ exudation of phlorotannins by the sublittoral kelp *Ecklonia radiata*. Mar Biol 121:349–354
Jensen PR, Harvell CD, Wirtz K, Fenical W (1996) Antimicrobial activity of extracts of Caribbean gorgonian corals. Mar Biol 125:411–419
Johannes RE (1967) Ecology of organic aggregates in the vicinity of a coral reef. Limnol Oceanogr 12:189–195
Joint I, Callow ME, Callow JA, Clarke KR (2000) The attachment of *Enteromorpha* zoospores to a bacterial biofilm assemblage. Biofouling 16:151–158

Kawamata M, Kon-ya K, Miki W (1994) Trigonelline, an antifouling substance isolated from an octocoral *Dendronephthya* sp. Fish Sci 60:485–486
Kelly SR, Jensen PR, Henkel TP, Fenical W, Pawlik JR (2003) Effects of Caribbean sponge extracts on bacterial attachment. Aquat Microb Ecol 31:175–182
Keough MJ (1984) Dynamics of the epifauna of the bivalve *Pinna bicolor*: interactions among recruitment, predation, and competition. Ecology 65:667–688
Keough MJ (1989) Kin recognition and the spatial distribution of larvae of the bryozoan *Bugula neritina*. Evolution 38:142–147
Keough MJ, Raimondi PT (1995) Responses of settling invertebrate larvae to bioorganic films: effects of different types of films. J Exp Mar Biol Ecol 185:235–253
Keifer PA, Rinehart KL, Hooper IR (1986) Renillafoulins, antifouling diterpenes from the sea pansy *Renilla reniformis* (Octocorallia). J Org Chem 51:4450–4454
Keifer PA, Schwartz RE, Koker MES, Hughes RG, Rittschof D, Rinehart KL (1991) Bioactive bromopyrrole metabolites from the Caribbean sponge *Agelas conifera*. J Org Chem 56:2965–2975
Keough MJ, Raimondi PT (1996) Responses of settling invertebrate larvae to bioorganic films: effects of large-scale variation in films. J Exp Mar Biol Ecol 207:59–78
Kirchman D, Graham S, Reish D, Mitchell R (1982) Lectins may mediate in the settlement and metamorphosis of *Janua (Dexiospira) brasiliensis* Grube (Polychaeta: Spirorbidae). Mar Biol Lett 3:131–142
Koh EGL, Sweatman H (2000) Chemical warfare among scleractinians: bioactive natural products from *Tubastraea faulkneri* Wells kill larvae of potential competitors. J Exp Mar Biol Ecol 251:141–160
Koh EGL, Tan TK, Chou LM, Goh NKC (2002) Antifungal properties of Singapore gorgonians: a preliminary study. J Exp Mar Biol Ecol 273:121–130
Kon-ya K, Shimidzu N, Adachi K, Miki W (1994a) 2,5,6-Tribromo-1-methylgramine, an antifouling substance from the marine bryozoan *Zoobrotryon pellucidum*. Fish Sci 60:773–775
Kon-ya K, Shimidzu N, Miki W, Endo M (1994b) Indole derivatives as potent inhibitors of larval settlement by the barnacle, *Balanus amphitrite*. Biosci Biotechnol Biochem 58:2178–2181
Kon-ya K, Shimidzu N, Otaki N, Yokoyama A, Adachi K, Miki W (1995) Inhibitory effect of bacterial ubiquinones on the settling of barnacle, *Balanus amphitrite*. Experientia 51:153–155
König GM, Wright AD (1997) *Laurencia rigida*: chemical investigations of its antifouling dichloromethane extract. J Nat Prod 60:967–970
Krug PJ, Zimmer RK (2000) Developmental dimorphism and expression of chemosensory-mediated behavior: Habitat selection by a specialist marine herbivore. J Exp Biol 203:1741–1754
Krug PJ, Boyd KG, Faulkner DJ (1995) Isolation and synthesis of tanyolides A and B, metabolites of the nudibranch *Sclerodoris tanya*. Tetrahedron 51:11063–11074
Kubanek J, Whalen KE, Engel S, Kelly SR, Henkel TP, Fenical W, Pawlik JR (2002) Multiple defensive roles for triterpene glycosides from two Caribbean sponges. Oecologia 131:125–136
Lam C, Harder T, Qian PY (2003) Induction of larval settlement in the polychaete *Hydroides elegans* by surface-associated settlement cues of marine benthic diatoms. Mar Ecol Prog Ser 263:83–92
Lau SCK, Mak KKW, Chen F, Qian PY (2002) Bioactivity of bacterial strains isolated from marine biofilms in Hong Kong waters for the induction of larval settlement in the marine polychaete *Hydroides elegans*. Mar Ecol Prog Ser 226:301–310
Lau SCK, Thiyagarajan V, Qian PY (2003) The bioactivity of bacterial isolates in Hong Kong waters for the inhibition of barnacle (*Balanus amphitrite* Darwin) settlement. J Exp Mar Biol Ecol 282:43–60
Laudien J, Wahl M (1999) Indirect effects of epibiosis on host mortality: seastar predation on differently fouled mussels. Mar Ecol 20:35–47

Lee OO, Qian PY (2003) Chemical control of bacterial epibiosis and larval settlement of *Hydroides elegans* in the red sponge *Mycale adherens*. Biofouling 19:171–180
Leitz T, Wagner T (1993) The marine bacterium *Alteromonas espejiana* induces metamorphosis of the hydroid *Hydractinia echinata*. Mar Biol 115:173–178
Littler MM, Littler DS (1999) Blade abandonment/proliferation: a novel mechanism for rapid epiphyte control in marine macrophytes. Ecology 80:1736–1746
Lopanik N, Lindquist N, Targett N (2004) Potent cytotoxins produced by a microbial symbiont protect host larvae from predation. Oecologia 139:131–139
Mafra-Neto A, Cardé RT (1994) Fine-scale structure of pheromone plumes modulates upwind orientation of flying moths. Nature 369:142–144
Maida M, Sammarco PW, Coll JC (1995) Effects of soft corals on scleractinian coral recruitment. I. Directional allelopathy and inhibition of settlement. Mar Ecol Prog Ser 121:191–202
Maki J, Rittschof D, Costlow J, Mitchell R (1988) Inhibition of attachment of larval barnacles, *Balanus amphitrite*, by bacterial biofilms. Mar Biol 97:199–206
Maldonado M, Young CM (1999) Effects of the duration of larval life on postlarval stages of the demosponge *Sigmadocia caerulea*. J Exp Mar Biol Ecol 232:9–21
Manning LM, Lindquist N (2003) Helpful habitant or pernicious passenger: interactions between an infaunal bivalve, an epifaunal hydroid and three potential predators. Oecologia 134:415–422
Maximilien R, de Nys R, Holmstrom C, Gram L, Givskov M, Crass K, Kjelleberg S, Steinberg PD (1998) Chemical mediation of bacterial surface colonisation by secondary metabolites from the red alga *Delisea pulchra*. Aquat Microb Ecol 15:233–246
McClintock JB, Baker BJ, Hamann MT, Yoshida W, Slattery M, Heine JN, Bryan PJ, Jayatilake GS, Moon BH (1994) Homarine as a feeding deterrent in common shallow-water Antarctic lamellarian gastropod *Marseniopsis mollis* – a rare example of chemical defense in a marine prosobranch. J Chem Ecol 20:2539–2549
Meadows PS, Williams GB (1963) Settlement of *Spirorbis borealis* Daudin larvae on surfaces bearing films of micro-organisms. Nature 198:610–611
Mearns-Spragg A, Bregu M, Boyd KG, Burgess JG (1998) Cross-species induction and enhancement of anti-microbial activity produced by epiphytic bacteria from marine algae and invertebrates, after exposure to terrestrial bacteria. Lett Appl Microbiol 27:142–146
Mihm JW, Banta WC, Loeb GI (1981) Effects of adsorbed organic and primary fouling films on bryozoan settlement. J Exp Mar Biol Ecol 54:167–179
Minchin D, Gollasch S (2003) Fouling and ships' hulls: how changing circumstances and spawning events may result in the spread of exotic species. Biofouling 19:111–122
Morse DE, Hooker N, Duncan H, Jensen L (1979) γ-Aminobutyric acid, a neurotransmitter, induces planktonic abalone larvae to settle and begin metamorphosis. Science 204: 407–410
Mullineaux LS, Butman CA (1991) Initial contact, exploration, and attachment of barnacle cyprids settling in flow. Mar Biol 110:93–103
Nakao Y, Masuda A, Matsunaga S, Fusetani N (1999) Pseudotheonamides, serine protease inhibitors from the marine sponge *Theonella swinhoei*. J Am Chem Soc 121:2425–2431
Nakao Y, Fujita M, Warabi K, Matsunaga S, Fusetani N (2000) Bioactive marine metabolites, part 104. Miraziridine A, a novel cysteine protease inhibitor from the marine sponge *Theonella aff. mirabilis*. J Am Chem Soc 122:10462–10463
Neal AL, Yule AB (1994a) The interaction between *Elminius modestus* Darwin cyprids and biofilms of *Deleya marina* NCMB1877. J Exp Mar Biol Ecol 176:123–139
Neal AL, Yule AB (1994b) The tenacity of *Elminius modestus* and *Balanus perforatus* cyprids to bacterial films grown under different shear regimes. J Mar Biol Assoc UK 74:251–257

Negri AP, Webster NS, Hill RT, Heyward AJ (2001) Metamorphosis of broadcast spawning corals in response to bacteria isolated from crustose algae. Mar Ecol Prog Ser 223:121–131
Netherton JC, Gurin S (1982) Biosynthesis and physiological role of homarine in marine shrimp. J Biol Chem 257:1971–1975
Newbold RW, Jensen PR, Fenical W, Pawlik JR (1999) Antimicrobial activity of Caribbean sponge extracts. Aquat Microb Ecol 19:279–284
Nevitt GA, Veit RR, Karieva PM (1995) Dimethyl sulfide as a foraging cue for Antarctic Procellariiform seabirds. Nature 376:680–682
Ng TY-T, Keough MJ (2003) Delayed effects of larval exposure to Cu in the bryozoan *Watersipora subtorquata*. Mar Ecol Prog Ser 257:77–85
Okino T, Yoshimura E, Hirota H, Fusetani N (1995) Antifouling kalihinenes from the marine sponge *Acanthella cavernosa*. Tetrahedron Lett 36:8637–8640
Okino T, Yoshimura E, Hirota H, Fusetani N (1996a) New antifouling kalihipyrans from the marine sponge *Acanthella cavernosa*. J Nat Prod 59:1081–1083
Okino T, Yoshimura E, Hirota H, Fusetani N (1996b) New antifouling sesquiterpenes from four nudibranchs of the family Phyllidiidae. Tetrahedron 52:9447–9454
Okuda RK, Scheuer PJ, Hochlowski JE, Walker RP, Faulkner DJ (1983) Sesquiterpenoid constituents of eight porostome nudibranchs. J Org Chem 48:1866–1869
O'Toole G, Kolter R (1998) Flagellar and twitching motility are necessary for *Pseudomonas aeruginosa* biofilm development. Mol Microbiol 30:295–304
Paul J, Jeffrey W (1985) Evidence for separate adhesion mechanisms for hydrophilic and hydrophobic surfaces in *Vibrio proteolytica*. Appl Environ Microbiol 50:431–437
Pawlik J (1992) Chemical ecology of the settlement of benthic marine invertebrates. Oceanogr Mar Biol Annu Rev 30:273–335
Pawlik JR (1993) Marine invertebrate chemical defenses. Chem Rev 93:1911–1922
Pawlik JR, Butman CA (1993) Settlement of a marine tube worm as a function of current velocity – interacting effects of hydrodynamics and behavior. Limnol Oceanogr 38:1730–1740
Pettit GR, Herald CL, Doubek DL, Herald DL (1982) Isolation and structure of bryostatin 1. J Am Chem Soc 104:6846–6848
Pettit GR, Kamano Y, Aoyagi R, Herald CL, Doubek DL, Schmidt JM, Rudloe JJ (1985) Antineoplastic agents 100: the marine bryozoan *Amathia convoluta*. Tetrahedron 41:985–994
Polychronopoulos P, Magiatis P, Skaltsounis AL, Tillequin F, Vardala-Theodorou E, Tsarbopoulos A (2001) Homarine, a common metabolite in edible Mediterranean molluscs: Occurrence, spectral data and revision of a related structure. Nat Prod Lett 15:411–418
Qian PY, Thiyagarajan V, Lau SCK, Cheung SCK (2003) Relationship between bacterial community profile in biofilm and attachment of the acorn barnacle *Balanus amphitrite*. Aquat Microb Ecol 33:225–237
Raimondi PT, Keough MJ (1990) Behavioral variability in marine larvae. Aust J Ecol 15:427–437
Renner MK, Shen YC, Cheng XC, Jensen PR, Frankmoelle W, Kauffman CA, Fenical W, Lobkovsky E, Clardy J (1999) Cyclomarins A-C, new antiinflammatory cyclic peptides produced by a marine bacterium (*Streptomyces* sp.). J Am Chem Soc 121:11273–11276
Rinehart KL, Kobayashi J, Harbour GC, Gilmore J, Mascal M, Holt TG, Shield LS, Lafargue F (1987) Eudistomins A-Q, beta carbolines from the antiviral Caribbean tunicate *Eudistoma olivaceum*. J Am Chem Soc 109:3378–3387
Rittschof D, Hooper IR, Costlow JD (1988) Settlement inhibition of marine invertebrate larvae: comparison of sensitivities of bryozoan and barnacle larvae. In: Thompson MF, Sarojini R, Nagabhushanam R (eds) Marine biodeterioration. Oxford and IBJ Publ, New Delhi, pp 599–608
Rohwer F, Seguritan V, Azam F, Knowlton N (2002) Diversity and distribution of coral-associated bacteria. Mar Ecol Prog Ser 243:1–10

Roughgarden J, Gaines S, Possingham H (1988) Recruitment dynamics in complex life cycles. Science 241:1460–1466

Russell G, Veltkamp CJ (1984) Epiphyte survival on skin-shedding macrophytes. Mar Ecol Prog Ser 18:149–153

Sand-Jensen K, Revsbech E (1987) Photosynthesis and light adaptation in epiphyte-macrophyte associations measured by oxygen microelectrodes. Limnol Oceanogr 32:452–457

Sata NU, Sugano M, Matsunaga S, Fusetani N (1999) Bioactive marine metabolites, part 88. Sinulamide: an H,K-ATPase inhibitor from a soft coral *Sinularia* sp. Tetrahedron Lett 40:719–722

Sato A, Fenical W (1983) Gramine-derived bromo-alkaloids from the marine bryozoan *Zoobotryon verticillatum*. Tetrahedron Lett 24:481–484

Sauer K, Camper A, Erhlich G, Costerton J, Davies D (2002) *Pseudomonas aeruginosa* displays multiple phenotypes during development as a biofilm. J Bacteriol 184:1140–1154

Scardino A, de Nys R, Ison O, O'Connor W, Steinberg PD (2003) Microtopography and antifouling properties of the shell surface of the bivalve molluscs *Mytilus galloprovincialis* and *Pincada imbricata*. Biofouling 19:221–230

Schmitt TM, Hay ME, Lindquist N (1995) Constraints on chemically mediated coevolution – multiple functions for seaweed secondary metabolites. Ecology 76:107–123

Schmitt TM, Lindquist N, Hay ME (1998) Seaweed secondary metabolites as antifoulants: effects of *Dictyota* spp. diterpenes on survivorship, settlement and development of marine invertebrate larvae. Chemoecology 8:125–131

Schulte B, de Nys R, Bakus G, Crews P, Eid C, Naylor S, Manes L (1991) A modified allomone collecting apparatus. J Chem Ecol 17:1327–1332

Sears MA, Gerhart DJ, Rittschof D (1990) Antifouling agents from the marine sponge *Lissodendoryx isodictyalis* Carter. J Chem Ecol 16:791–799

Shinagawa A, Suzuki T, Konosu S (1995) Preliminary studies on the effects of salinity on intracellular nitrogenous osmolytes in various tissues and hemolymph of the Japanese spiny lobster, *Panulirus japonicus* (Vonsiebold, 1824). Crustaceana 68:129–137

Sieburth J, Tootle J (1981) Seasonality of microbial fouling on *Ascophyllum nodosum* (L.) Lejol, *Fucus vesiculosus* L., *Polysiphonia lanosa* (L.) Tandy and *Chondrus crispus* Stackh. J Phycol 17:57–64

Sjögren M, Goransson U, Johnson AL, Dahlström M, Andersson R, Bergman J, Jonsson PR, Bohlin L (2004) Antifouling activity of brominated cyclopeptides from the marine sponge *Geodia barretti*. J Nat Prod 67:368–372

Slattery M (1994) A comparative study of population structure and chemical defenses in the soft corals *Alcyonium paessleri* May, *Clavularia frankliniana* Rouel, and *Gersemia antarctica* Kukenthal in McMurdo Sound, Antarctica. PhD Thesis, University of Alabama at Birmingham, 109 pp

Slattery M, McClintock JB, Heine JN (1995) Chemical defenses in Antarctic soft corals: evidence for antifouling compounds. J Exp Mar Biol Ecol 190:61–77

Slattery M, Hamann MT, McClintock JB, Perry TL, Puglisi MP, Yoshida WY (1997) Ecological roles for water-borne metabolites from Antarctic soft corals. Mar Ecol Prog Ser 161:133–144

Stachowicz JJ, Hay ME (1999) Reducing predation through chemically mediated camouflage: indirect effects of plant defenses on herbivores. Ecology 80:495–509

Standing J, Hooper IR, Costlow JD (1984) Inhibition and induction of barnacle settlement by natural products present in octocorals. J Chem Ecol 10:823–834

Steinberg PD, de Nys R, Kjelleberg S (1998) Chemical inhibition of epibiota by Australian seaweeds. Biofouling 12:227–244

Steinberg PD, de Nys R, Kjelleberg S (2001) Chemical mediation of surface colonization. In: McClintock J, Baker B (eds) Marine chemical ecology. CRC Press, London, pp 355–387

Stoecker D (1978) Resistance of a tunicate to fouling. Biol Bull 155:615–626

Stoecker D (1980) Relationships between chemical defense and ecology in benthic ascidians. Mar Ecol Prog Ser 3:257–265

Strathmann R, Branscomb E, Vedder K (1981) Fatal errors in set as a cost of dispersal and the influence of intertidal flora on set of barnacles. Oecologia 48:13–18

Sutherland IW (2001) The biofilm matrix – an immobilized but dynamic microbial environment. Trends Microbiol 9:222–227

Szewzyk U, Holmström C, Wrangstadh M, Samuelsson MO, Maki JS, Kjelleberg S (1991) Relevance of the exopolysaccharide of marine *Pseudomonas* sp. strain S9 for the attachment of *Ciona intestinalis* Larvae. Mar Ecol Prog Ser 75:259–265

Takewati M, Provasoli L, Pinter IJ (1983) Morphogenesis of *Monostroma oxysperma* (Kutz.) Doty (Chlorophycea) in axenic culture, especially in bi-algal culture. J Phycol 19:404–416

Tamburri MN, Zimmer-Faust RK (1996) Suspension feeding: basic mechanisms controlling recognition and ingestion of larvae. Limnol Oceanogr 41:1188–1197

Tamburri MN, Zimmer-Faust RK, Tamplin ML (1992) Natural sources and properties of chemical inducers mediating settlement of oyster larvae – a reexamination. Biol Bull 183:327–338

Tamburri MN, Finelli CM, Wethey DS, Zimmer-Faust RK (1996) Chemical induction of larval settlement behavior in flow. Biol Bull 191:367–373

Tapiolas DM, Roman M, Fenical W, Stout TJ, Clardy J (1991) Octalactin A and octalactin B: cytotoxic 8-membered-ring lactones from a marine bacterium, *Streptomyces* sp. J Am Chem Soc 113:4682–4683

Targett NM, Bishop SS, McConnell OJ, Yoder JA (1983) Antifouling agents against the benthic marine diatom, *Navicula salinicola* - homarine from the gorgonians *Leptogorgia virgulata* and *L. setacea* and analogs. J Chem Ecol 9:817–829

Tarjuelo I, Lopez-Legentil S, Codina M, Turon X (2002) Defense mechanisms of adults and larvae of colonial ascidians: patterns of palatability and toxicity. Mar Ecol Prog Ser 235:103–115

Taylor EM, Zheng D, Lee M, Troy PJ, Gyananath G, Sharma SK (1997) Influence of surface properties on accumulation of conditioning films and marine bacteria on substrata exposed to oligotrophic waters. Biofouling 11:31–57

Taylor MW, Schupp PJ, Baillie HJ, Charlton TS, de Nys R, Kjelleberg S, Steinberg PD (2004a) Evidence for acyl homoserine lactone signal production in bacteria associated with marine sponges. Appl Environ Microbiol 70:4387–4389

Taylor MW, Schupp PJ, Dahllof I, Kjelleberg S, Steinberg PD (2004b) Host specificity in marine sponge-associated bacteria, and potential implications for marine microbial diversity. Environ Microbiol 6:121–130

Thomason JC, Hills JM, Mapson P (2000) The consequences of seasonal reproductive strategies for the interpretation of settlement trials. Biofouling 16:323–329

Thomason JC, Hills JM, Thomason PO (2002) Field-based behavioural bioassays for testing the efficacy of antifouling coatings. Biofouling 18:285–292

Thompson JE (1985) Exudation of biologically active metabolites in the sponge *Aplysina fistularis*. 1. Biological evidence. Mar Biol 88:23–26

Thompson J, Barrow K, Faulkner D (1983) Localization of two brominated metabolites, aerothionin and homoaerothionin, in spherulous cells of the marine sponge *Aplysina fistularis* (= *Verongia thiona*). Acta Zool 64:199–210

Thompson JE, Walker RP, Faulkner DJ (1985) Screening and bioassays for biologically-active substances from 40 marine sponge species from San Diego, California, USA. Mar Biol 88:11–21

Thompson RC, Norton TA, Hawkins SJ (1998) The influence of epilithic microbial films on the settlement of *Semibalanus balanoides* cyprids – a comparison between laboratory and field experiments. Hydrobiologia 376:203–216

Tomono Y, Hirota H, Fusetani N (1999) Isogosterones A-D, antifouling 13,17-*seco*steroids from an octocoral *Dendronephthya* sp. J Org Chem 64:2272–2275

Toonen RJ, Pawlik JR (2001) Foundations of gregariousness: a dispersal polymorphism among the planktonic larvae of a marine invertebrate. Evolution 55:2439–2454

Tosuji H, Fusetani N, Seki Y (2003) Calyculin A causes the activation of histone H-1 kinase and condensation of chromosomes in unfertilized sea urchin eggs independently of the maturation-promoting factor. Comp Biochem Physiol 135C:415–424

Trischman JA, Tapiolas DM, Jensen PR, Dwight R, Fenical W, McKee TC, Ireland CM, Stout TJ, Clardy J (1994) Salinamide A and salinamide B: anti-inflammatory depsipeptides from a marine streptomycete. J Am Chem Soc 116:757–758

Tsukamoto S, Kato H, Hirota H, Fusetani N (1996a) Mauritiamine, a new antifouling oroidin dimer from the marine sponge *Agelas mauritiana*. J Nat Prod 59:501–503

Tsukamoto S, Kato H, Hirota H, Fusetani N (1996b) Ceratinamides A and B: new antifouling dibromotyrosine derivatives from the marine sponge *Pseudoceratina purpurea*. Tetrahedron 52:8181–8186

Tsukamoto S, Kato H, Hirota H, Fusetani N (1997) Antifouling terpenes and steroids against barnacle larvae from marine sponges. Biofouling 11:283–291

Turner EJ, Zimmer-Faust RK, Palmer MA, Luckenback M, Pentcheff ND (1994) Settlement of oyster (*Crassostrea virginica*) larvae: effects of water flow and a water-soluble chemical cue. Limnol Oceanogr 39:1579–1593

Unabia CRC, Hadfield MG (1999) Role of bacteria in larval settlement and metamorphosis of the polychaete *Hydroides elegans*. Mar Biol 133:55–64

Unson MD, Faulkner DJ (1993) Cyanobacterial symbiont biosynthesis of chlorinated metabolites from *Dysidea herbacea* (Porifera). Experientia 49:349–353

Unson MD, Holland ND, Faulkner DJ (1994) A brominated secondary metabolite synthesized by the cyanobacterial symbiont of a marine sponge and accumulation of the crystalline metabolite in the sponge tissue. Mar Biol 119:1–11

Uriz M, Martin D, Turon X, Ballesteros E, Hughes R, Acebal C (1991) An approach to the ecological significance of chemically mediated bioactivity in Mediterranean benthic communities. Mar Ecol Prog Ser 70:75–88

Uriz M, Becerro M, Tur J, Turon X (1996) Location of toxicity within the Mediterranean sponge *Crambe crambe* (Demonspongiae: Poecilosclerida). Mar Biol 124:583–590

Vance RR (1978) A mutualistic interaction between a sessile marine clam and its epibionts. Ecology 59:679–685

Vickers NJ, Baker TC (1994) Reiterative responses to single strands of odor promote sustained upwind flight and odor source location by moths. Proc Natl Acad Sci USA 91:5756–5760

Wahl M (1989) Marine epibiosis. 1. Fouling and antifouling – some basic aspects. Mar Ecol Prog Ser 58:175–189

Wahl M (1995) Bacterial epibiosis on Bahamian and Pacific ascidians. J Exp Mar Biol Ecol 191:239–255

Wahl M (1996) Fouled snails in flow: potential of epibionts on *Littorina littorea* to increase drag and reduce snail growth rates. Mar Ecol Prog Ser 138:157–168

Wahl M (1997) Increased drag reduces growth of snails: comparison of flume and in situ experiments. Mar Ecol Prog Ser 151:291–293

Wahl M, Lafargue F (1990) Marine epibiosis. 2. Reduced fouling on *Polysyncraton lacazei* (Didemnidae, Tunicata) and proposal of an antifouling potential index. Oecologia 82:275–282

Wahl M, Banaigs B (1991) Marine epibiosis. 3. Possible antifouling defense adaptations in *Polysyncraton lacazei* (Giard) (Didemnidae, Ascidiacea). J Exp Mar Biol Ecol 145:49–63

Wahl M, Sonnichsen H (1992) Marine epibiosis. 4. The periwinkle *Littorina littorea* lacks typical antifouling defenses – why are some populations so little fouled? Mar Ecol Prog Ser 88:225–235

Wahl M, Jensen PR, Fenical W (1994) Chemical control of bacterial epibiosis on ascidians. Mar Ecol Prog Ser 110:45–57

Wahl M, Kroger K, Lenz M (1998) Non-toxic protection against epibiosis. Biofouling 12:205–226

Walker G, Yule AB (1984) Temporary adhesion of the barnacle cyprid: the existence of an antennular adhesive secretion. J Mar Biol Assoc UK 64:679–686

Walker RP, Thompson JE, Faulkner DJ (1985) Exudation of biologically active metabolites in the sponge *Aplysina fistularis*. 2. Chemical evidence. Mar Biol 88:27–32
Walls J, Ritz D, Blackman A (1993) Fouling, surface bacteria, and antibacterial agents of four bryozoan species from in Tasmania, Australia. J Exp Mar Biol Ecol 169:1–13
Warabi K, Matsunaga S, van Soest RWM, Fusetani N (2003) Dictyodendrins A-E, the first telomerase-inhibitory marine natural products from the sponge *Dictyodendrilla verongiformis*. J Org Chem 68:2765–2770
Weissburg MJ, Zimmer-Faust RK (1994) Odor plumes and how blue crabs use them in finding prey. J Exp Biol 197:349–375
Whitely M, Bangera M, Bumgarner R, Parsek M, Teitzel G, Lory S, Greenberg E (2001) Gene expression in *Pseudomonas aeruginosa* biofilms. Nature 413:860–864
Wicksten MK (1983) Camouflage in marine invertebrates. Oceanogr Mar Biol Annu Rev 21:177–193
Witman JD, Suchanek TH (1984) Mussels in flow: drag and dislodgement by epizoans. Mar Ecol Prog Ser 16:259–268
Whitchurch CB, Tolker-Nelson T, Ragas PC, Mattick JS (2002) Extracellular DNA required for bacterial biofilm formation. Science 295:1487
Wieczorek SK, Clare AS, Todd CD (1995) Inhibitory and facilitatory effects of microbial films on settlement of *Balanus amphitrite amphitrite* larvae. Mar Ecol Prog Ser 119:221–228
Wieczorek SK, Todd CD (1997) Inhibition and facilitation of bryozoan and ascidian settlement by natural multi-species biofilms: effects of film age and the roles of active and passive larval attachment. Mar Biol 128:463–473
Wigglesworth-Cooksey B, Cooksey KE (1996) A computer-based image analysis system for biocide screening. Biofouling 10:225–237
Wikström SA, Pavia H (2004) Chemical settlement inhibition versus post-settlement mortality as an explanation for differential fouling of two congeneric seaweeds. Oecologia 138:223–230
Wilsanand V, Wagh AB, Bapuji M (2001) Antifouling activities of octocorals on some microfoulers. Microbios 104:131–140
Willemsen PR, Overbeke K, Suurmond A (1998) Repetitive testing of TBTO, sea-nine 211 and farnesol using *Balanus amphitrite* (Darwin) cypris larvae: Variability in larval sensitivity. Biofouling 12:133–147
Woollacott RM, Hadfield MG (1996) Induction of metamorphosis in larvae of a sponge. Invert Biol 115:257–262
Yan L, Boyd KG, Burgess JG (2002) Surface attachment induced production of antimicrobial compounds by marine epiphytic bacteria using modified roller bottle cultivation. Mar Biotechnol 4:356–366
Yan L, Boyd KG, Adams DR, Burgess JG (2003) Biofilm-specific cross-species induction of antimicrobial compounds in bacilli. Appl Environ Microbiol 69:3719–3727
Zhao B, Qian PY (2002) Larval settlement and metamorphosis in the slipper limpet *Crepidula onyx* (Sowerby) in response to conspecific cues and the cues from biofilm. J Exp Mar Biol Ecol 269:39–51
Zimmer RK, Butman CA (2000) Chemical signaling processes in the marine environment. Biol Bull 198:168–187

Furanones

R. de Nys, M. Givskov, N. Kumar, S. Kjelleberg, P.D. Steinberg

Abstract. The red alga *Delisea pulchra* has been a model organism for understanding the ecological role of secondary metabolites as natural antifoulants. Furanones are produced by the plant and delivered to the surface at a concentration where they regulate bacterial colonisation and the settlement of epibiota. This biological understanding has led to the application of furanones as inhibitors of bacterial- and macro-fouling. Furanones inhibit bacterial colonisation and biofilm development through interference with a key bacterial quorum-sensing pathway, the acylated homoserine lactone regulatory system in Gram-negative bacteria. They also interfere with the alternative AI-2 signalling system in Gram-negative and Gram-positive bacteria. Synthetic programs have developed a library of more than 200 furanone and furanone-analogues including surface attached-furanones. These furanone analogues are potent anti-infectives and inhibit pathogenic phenotypes in Gram-negative and Gram-positive bacteria as demonstrated *in-vitro* using gene microarrays, and *in-vivo* using mouse models. Additionally, furanones inhibit the expression of bacterial exo-enzymes that actively degrade components of the immune system thereby enhancing the immune response. Surface-attached furanones immobilised on catheters also inhibit bacterial attachment and retain activity for extended periods. Furanones are strong deterrents of the settlement and growth of macrofouling organisms and as such have potential application as a marine antifouling technology. Laboratory antifouling assays have been used to identify effective and safe furanone-

R. de Nys
School of Marine Biology and Aquaculture, James Cook University, Townsville, Q4811, Australia

M. Givskov
Centre for Biomedical Microbiology, BioCentrum, Bldg 301, Technical University of Denmark, 2800 Lyngby, Denmark

N. Kumar
School of Chemistry, University of NSW, Sydney, NSW 2052, Australia

S. Kjelleberg
School of Biotechnology and Biomolecular Sciences and Centre for Marine Biofouling and Bio-innovation, University of New South Wales, Sydney, NSW 2052, Australia

P.D. Steinberg
School of Biological, Earth, and Environmental Science and Centre for Marine Biofouling and Bio-innovation, University of New South Wales, Sydney, NSW 2052, Australia

Progress in Molecular and Subcellular Biology
Subseries Marine Molecular Biotechnology
N. Fusetani, A.S. Clare (Eds.): Antifouling Compounds

analogues while field trials of furanones incorporated into coatings and polymers demonstrate efficacies similar to commercial biocides. Further development is required to control the release of compounds from suitable carriers to extend coating/polymer lifespans. This review summarises the extensive work on furanones focusing on their natural and applied antifouling activities.

1 Natural Furanones from *Delisea*

The colonisation and growth of marine organisms on natural living surfaces (seaweeds, invertebrates, etc.) can have a significant negative impact on the host organism. This is particularly the case for seaweeds as they grow in the photic zone where fouling pressure is highest. Host seaweeds can be damaged by both micro- and macro-foulers. Bacteria, viruses, fungi and microalgae (microfouling) can cause deformation, necrosis or tissue death in marine algae, with widespread destruction of algae by both bacteria and fungi (reviewed by Bouarab et al. 2001; Potin et al. 2002). Algal and invertebrate epibiota can also have significant negative impacts on seaweed hosts by decreasing photosynthesis, decreasing reproduction, and increasing the risk of removal from the substratum (reviewed by de Nys and Steinberg 1999). In some cases, biofouling can benefit host plants by decreasing herbivory (Wahl and Hay 1995), and reducing susceptibility to extremes of temperature, light and desiccation (Oswald and Seed 1986; de Nys and Steinberg 1999). However, evidence supporting the broad-spectrum benefits of fouling is rare. Therefore, a chemical defence against pathogenic bacteria, fungi, viruses, and/or epiphytic plants and animals has the potential to confer an ecological or evolutionary advantage to the host.

The red alga *Delisea pulchra* has been a model organism for understanding the ecological role of marine natural products in antifouling, and for the potential development of natural products into commercial products. *D. pulchra* (Greville) Montage (Bonnemaisonales, Bonnemaisoniaceae) (cf. *Delisea fimbriata*) (Millar 1990) is a common Australasian red alga that ranges from Southern Queensland to Tasmania in Australia and through the Antarctic Islands and Antarctic Peninsula (Bonin and Hawkes 1988). The genus *Delisea* produces a range of unique natural products – halogenated furanones or "fimbrolides", which have interesting biological activities in both a natural and applied context, and at both the ecological and molecular levels. *Delisea pulchra* produces more than 20 of these 2(5*H*)-furanones (Fig. 1) with structural variation based primarily on variable substitutions at two positions, in particular acetate and hydroxy groups at C-1' and halogens at C-6 (Kazlaukas et al. 1977; de Nys et al. 1993). The majority of these compounds are present as *E* and *Z* isomers (C-6) with the *Z* isomer being more abundant. These 2(5*H*)-furanones have also been identified from other members of the genus, including *D. elegans* (New Zealand) (McCombs et al. 1988), and *D. fimbriata* (King George Island) (Cueto et al. 1997). *D. elegans* also produces a range of novel furanones containing cyclobutane functions (McCombs et al. 1988).

The natural product chemistry of *D. pulchra* (hereafter referred to as *Delisea*) has also been studied quantitatively. While *Delisea* produces a broad array of furanones, four compounds (1–4) (Fig. 1) make up the majority (~95%) of the natural products chemistry of the plant and two furanones (3 and 4) (Fig. 1) are consistently the most abundant (de Nys et al. 1996, 1998; Dworjanyn et al. 1999; Rogers et al. 2000; Wright et al. 2004). There is significant quantitative variation in the concentration of these furanones within individual thalli (plants) (de Nys et al. 1996; Dworjanyn et al. 1999; Rogers et al. 2000) and between plants at both small (metres) and large geographic scales (Wright et al. 2000, 2004). Furanone concentrations also vary throughout the year (Wright et al. 2000) and between different life-history phases (Wright et al. 2004). Importantly, variation in the concentrations of total furanones, and the most abundant furanone (3), have a heritable component (Wright et al. 2004) indicating the potential for these compounds to evolve in response to selection from natural enemies, as predicted by models of chemical defence. These natural enemies potentially include herbivores (Williamson et al. 2004; Wright et al. 2004), and macro- and micro-fouling organisms (Maximilien et al. 1998; Dworjanyn 2001; Dworjanyn et al. 2006). The effects of furanones on fouling organisms are the focus of this review.

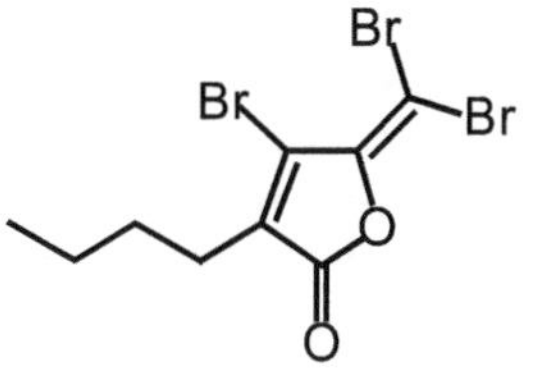

1

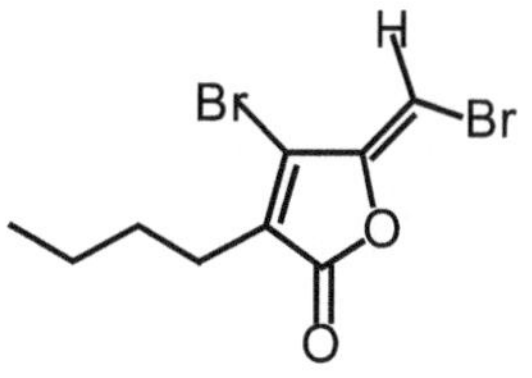

2

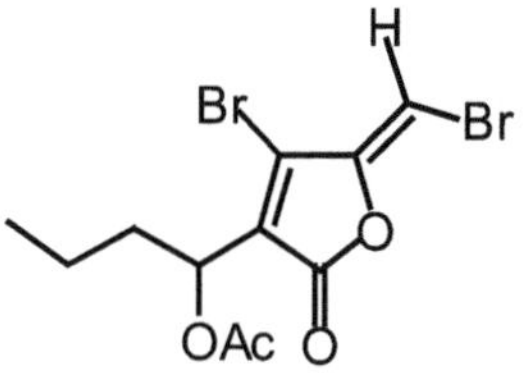

3

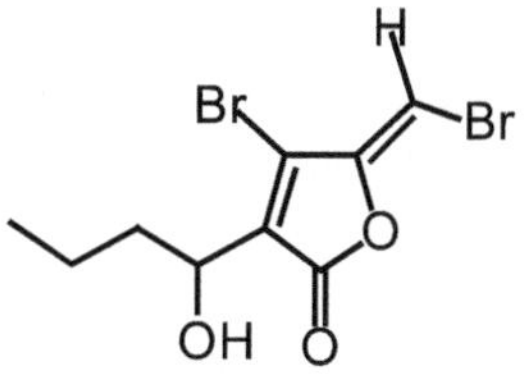

4

Fig. 1. Structures of natural furanones

2
Natural Antifouling Activity of Furanones

2.1
Surface Delivery and Surface Quantification of Furanones

The potential ecological and applied significance of furanones was recognised in the first chemical studies of *Delisea*. Pettus et al. (1977) noted that *Delisea* was unique in its environment (Antarctic Peninsula) because "it is remarkably free of normal fouling organisms (micro and macro) which colonize other algae in the immediate area". *Delisea* also stands out from co-occurring organisms in its sub-tidal in southeast Australia with consistently low levels of both micro- (Maximilien et al. 1998) and macro-fouling (Dworjanyn 2001; Nylund et al. 2006). These observations, and the strong biological effects of furanones from this alga, suggested that these compounds might play a role in keeping the alga relatively free from fouling. However, because antifouling is a surface mediated phenomenon, it was necessary to demonstrate the deployment of these compounds at the surface of the alga and their concentration on the surface where fouling organisms would contact "natural antifoulants".

The localisation and delivery of furanones in *Delisea* was determined using a combination of fluorescence microscopy, culture studies and quantitative analysis (Dworjanyn et al. 1999). Furanones are produced and contained within gland cells that are characterised by a lack of chloroplasts and a large central vesicle (Fig. 2). The furanones are localised within the vesicles and appear white-light blue (435 nm) when irradiated with UV light (360–410 nm) (Fig. 2) (Dworjanyn et al. 1999). Gland cells are present from the early development of the plant and are distributed throughout the thallus and on the surface of the plant (Fig. 3) (Dworjanyn et al. 1999). The density of cells, and consequently furanones, is significantly higher in the tip of the plant than the middle and base of *Delisea*. The presence of furanones in gland cells of *Delisea* was confirmed by culturing tetraspores in media with, and without, bromide, but identical in every other respect. Bromide is not essential for the growth of red algae (Fries 1966) but is an element in all furanones (Fig. 1). Gland cells were visible after one week of culture and the number and size of gland cells was not significantly different for plants grown in media with or without bromide. However, the vesicle within the gland cells of plant grown without bromide was smaller and furanones were not detectable in these plants either by high-resolution gas-chromatography or epi-fluorescence microscopy (Dworjanyn et al. 1999). Furanones were quantified in plants grown in media containing bromide and emission spectra of the furanones visualised by microscopy (Dworjanyn et al. 1999) demonstrated the localisation of furanones within the vesicles of gland cells.

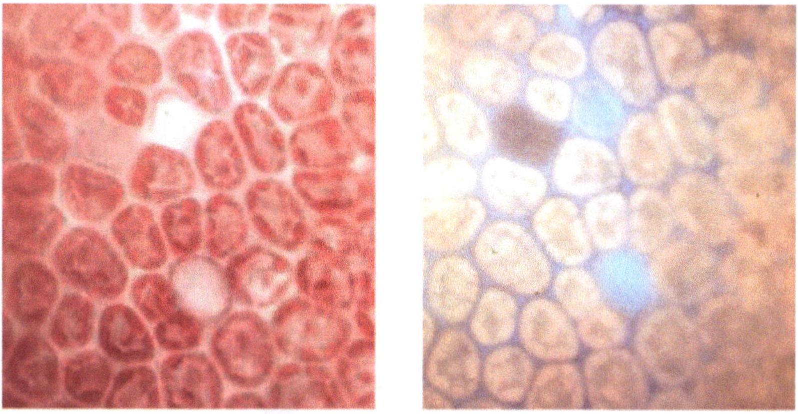

Fig. 2. **A** Light micrograph of a surface view of the thallus of *Delisea pulchra* showing gland cells surrounded by cortical cells, **B** fluorescence micrograph of the same surface view of the thallus of *Delisea pulchra*. The white-blue autofluorescence (435 nm) of furanones is visible in gland cells (photographs by Nicholas Paul)

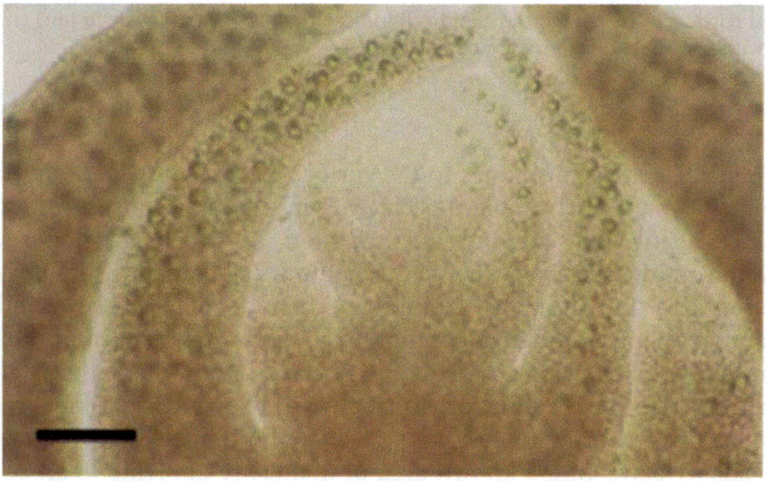

Fig. 3. Light micrograph of the apical tip of *Delisea pulchra* showing the high density of gland cells in the meristematic region of the plant (*scale bar* 100 µm) (reproduced with permission from Dworjanyn et al. 1999)

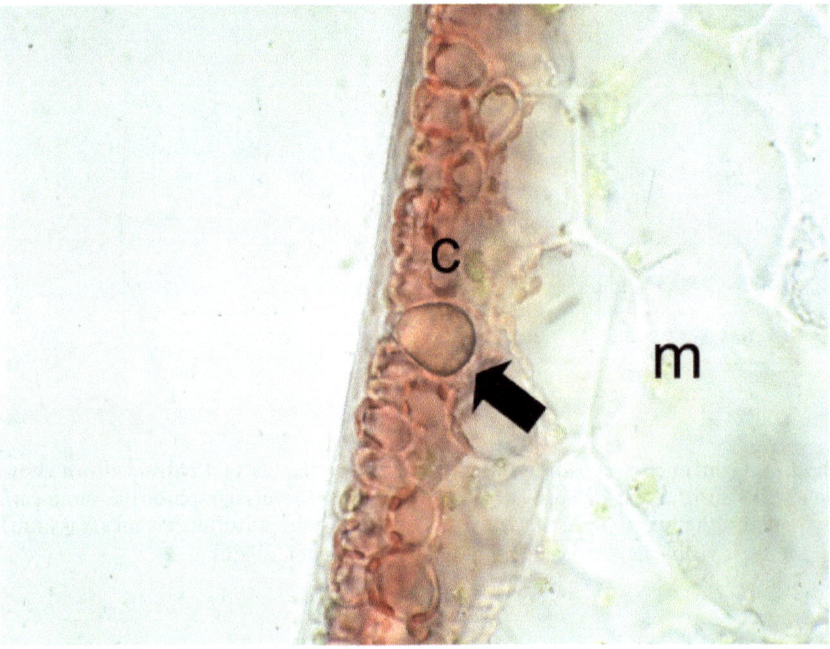

Fig. 4. Light micrograph of the transverse section of *Delisea pulchra* showing medullary (*m*) and cortical cells (*c*) with the gland cell indicated by arrow (*scale bar* 10 μm) (reproduced with permission from Dworjanyn et al. 1999)

Gland cells deliver furanones to the surface of the plant where they are able to act as natural antifoulants (Fig. 4). These compounds can then be quantified using the "hexane dip" method in which furanones are removed from the surface of the thallus without cell damage using hexane extraction (de Nys et al. 1998). It can then be determined whether furanones, either as individual compounds or as a total of all compounds, are present at a concentration on the surface of the plant that deters ecologically relevant fouling organisms. We also note that this same technique can be used to test the effect of furanones against fouling organisms by extracting a known area of thallus, re-applying to an equal surface area, and testing against fouling organisms in laboratory or field assays. Using these techniques, the surface presentation of furanones at biologically relevant concentrations has been unequivocally determined (de Nys et al. 1998; Dworjanyn et al. 1999).

Furanones are present on the surface of *Delisea* with total surface concentrations of the four major furanones (Fig. 1) ranging from 100 to 500 ng cm^{-2} (de Nys et al. 1998; Dworjanyn et al. 1999). The most abundant compounds in the plant are also the most abundant on the surface with a concentration range of 25 to 122.4 ng cm^{-2} for furanone **3** to less than 10 ng cm^{-2} for furanone **1** (de Nys et al. 1998). Furanones are also distributed unevenly over the surface of the thallus with significantly higher levels of furanones occurring on the tip of the thallus than on the base, again reflecting the distribution of furanones within the plant (Dworjanyn et al. 1999). The mechanism of delivery also suggests a constant release of furanones onto the surface with surface-bound metabolites being between 0.2 to 0.4% of total extracted furanones from the plant (Dworjanyn et al. 1999). As it is difficult to determine the extraction efficiency of the technique, this method probably represents a minimum concentration of furanones on the surface.

2.2 Bacterial Fouling

The abundance of bacteria on *Delisea* is lower than on co-occurring algae as measured by scanning electron microscopy. Furthermore, while the abundance of bacteria is similar across different regions of the thallus of co-occurring species, bacterial numbers on the tip of *Delisea* are more than an order of magnitude less than the base of the plant and on the tip of co-occurring species (Maximilien et al. 1998). This is inversely correlated to variation in the concentration of furanones along the length of the thallus. These patterns suggest that furanones inhibit bacterial fouling, and furanones were tested at natural concentrations against bacteria in the laboratory and in the field to confirm this result.

Concentrations of furanones found on the surface of *Delisea* deter the settlement and colonisation of ecologically relevant micro- and macro-fouling organisms. Furanones strongly inhibited the attachment of bacteria in field assays at a concentration of 100 ng cm^{-2} corresponding to the natural concentration on the surface of the plant (Maximilien et al. 1998). However, the crude extract of *Delisea* and furanones did not inhibit the growth of environmental strains of bacteria including those isolated from the surface of *Delisea*. Rather, crude extract and furanones specifically inhibited the attachment of bacteria and their ability to swarm and swim, characteristics central to the colonisation of a surface. Furanones acted differentially against strains associated with *Delisea* and those isolated from the environment. At non-growth inhibitory concentrations attachment was most affected in strains not associated with, or isolated from *Delisea*, while swarming and swimming were most affected in strains isolated from *Delisea*. By inhibiting the attachment of selected bacteria, and regulating their growth, furanones appear to mediate the composition of the bacterial community on the surface of *Delisea* (Maximilien et al. 1998).

These experiments were among the first to demonstrate the chemical mediation of marine plant-microbe interactions (Engel et al. 2002). They also linked, for the first time, furanones with bacterial quorum sensing, as regulated by the acylated homoserine lactone (AHL) regulatory system (Swift et al. 1996), as a potential explanation of how furanones inhibit bacterial phenotypes without significant growth effects. The *in situ* ecological role of furanones in plant-microbe interactions remains an area for further development, particularly with the application of molecular methods in quantifying microbial communities and their response to the environment (Dahllof 2002). In contrast, the use of *Delisea* and furanones as a biological model for the development of therapeutic inhibitors of biofilm formation has progressed rapidly. The *ex-situ* interaction between furanones and AHLs, and the development of new antimicrobial technologies based on furanones is reviewed in Sect. 3.

2.3 Macrofouling

In a similar manner to bacterial fouling, macrofouling on *Delisea* is significantly lower than on co-occurring species, in some case by an-order-of-magnitude (Nylund et al. 2006). The major fouling organisms on *Delisea* and co-occurring species are algae (Dworjanyn et al. 2006; Nylund et al. 2006), although bryozoans and tubeworms occur in low abundance.

As expected from these studies, and earlier field observations (Pettus et al. 1977), furanones strongly inhibit the settlement and growth of macrofouling organisms. Surface extracts of *Delisea* contain furanones as the major component (de Nys et al. 1998) and when tested in assays deter the settlement and growth of the most common epiphytes of algae in the sub-tidal habitat where *Delisea* is most abundant. The crude surface extract of *Delisea* at the natural concentration completely deters the settlement and growth of *Polysiphonia* sp. carpospores, *Ulva* sp. gametes, *Ceramium* sp. tetraspores, and *Ectocarpus siliculosus* gametes, all of which are abundant on co-occurring species (Dworjanyn 2001, Dworjanyn et al. 2006). Furthermore, the most abundant furanones (**2** and **3**, Fig. 1) and the mixture of the four most abundant furanones (**1**–**4**, Fig. 1), in the same ratio as on the surface of the plant, inhibit the settlement and growth of these four epiphytes (Dworjanyn et al. 2006). Furanone **3** is the most abundant furanone both in the plant (de Nys et al. 1996; Wright et al. 2000) and on the surface of the plant with concentrations of 25–122 ng cm^{-2} (de Nys et al. 1998; Dworjanyn et al. 1999) and inhibits the settlement and growth of *Ulva* gametes by 90% at 10 ng cm^{-2} and completely inhibits the growth of the four epiphytes at 100 ng cm^{-2}. The natural mix of furanones occurs on the surface of the plant at concentrations of 50–250 ng cm^{-2} (de Nys et al. 1998; Dworjanyn et al. 1999), and also strongly inhibits *Ulva* gametes at 10 ng cm^{-2}, *Ceramium* tetraspores and *Polysiphonia* carpospores at 100 ng cm^{-2}, and *Ectocarpus* gametes at 1 μg cm^{-2} (Dworjanyn et al. 1999, Dworjanyn et al. 2006).

Not all furanones necessarily act as natural antifoulants. For example, furanone **2** had a surface concentration of less than 10 ng cm^{-2} but only deters the settlement of epiphytes at 1μg cm^{-2}, ruling out a primary role as an antifouling agent for this compound (although it may act synergistically with other furanones). This compound specific response has also been demonstrated in an earlier study with furanone **3** having significantly higher levels of activity against the settlement and growth of *Ulva* sp. than furanones **4**, **2** and **1** (de Nys et al. 1995). The natural antifouling effects of furanones are not restricted to epiphytic algae with the surface extract of *Delisea* deterring the settlement of the common fouling bryozoan *Bugula neritina*. In contrast, extracts of twelve co-occurring species have little or no deterrent effect (Steinberg et al. 2001).

3 The Mode of Action of Furanones – Inhibition of Bacterial Signalling Systems

The molecular or physiological mode of action of furanones against macrofoulers is unknown. However, one of the most significant aspects of the biological activity of these compounds is their ability to interfere with bacterial signal-mediated quorum sensing systems, and thus interfere with a range of phenotypes used by bacteria to colonise (foul) surfaces. Bacteria organise themselves structurally through the synthesis of, and response to, intercellular signal compounds (for reviews see Dunny and Winans 1999). This cell-to-cell signalling is referred to as quorum sensing (QS) because it enables a given bacterial species to sense when a critical (i.e., quorate) population density has been reached (for example in the host) and in response activate expression of target genes required for succession (Fuqua et al. 1994). A diverse range of bacterial metabolites such as peptides, butyrolactones, palmitic acid methyl esters, quinones, and cyclic dipeptides are recognized as intercellular signals which are used to monitor population size.

The most intensely studied and extensively characterized intercellular signals are the *N*-acyl-L-homoserine lactones or AHLs (Whitehead et al. 2001). AHLs (Fig. 5) are produced by over 40 proteobacterial species in a diverse range of habitats. Most AHL-producing bacteria are associated with higher organisms through symbiotic or pathogenic interactions (for reviews see Eberl 1999; Parsek and Greenberg 2000; Camara et al. 2002).

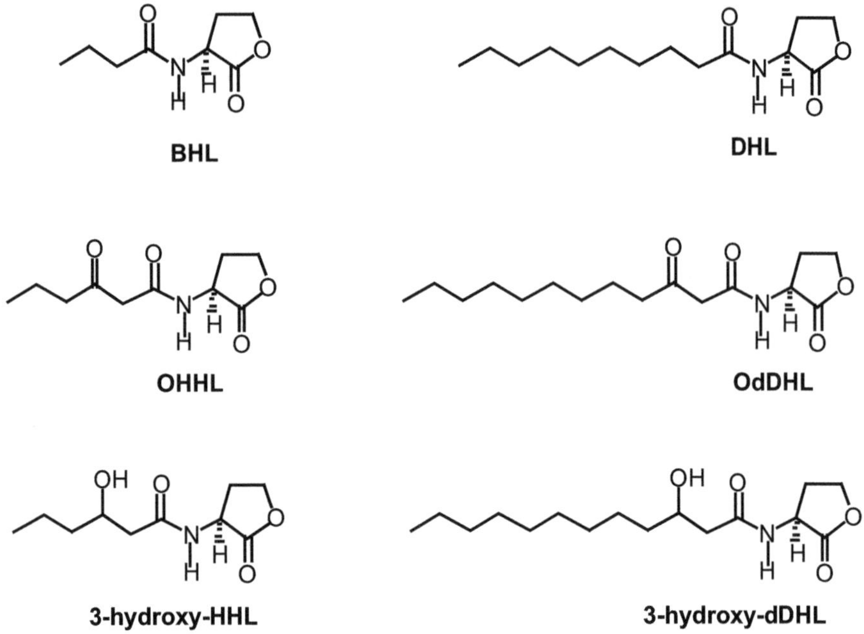

Fig. 5. Structures of representative classes of acylated homoserine lactones. *N*-butanoyl-L-homoserine lactone (*BHL*), *N*-decanoyl-L-homoserine lactone (*DHL*), *N*-3-(oxo-hexanoyl)-L-homoserine lactone (*OHHL*), *N*-3-(oxo-dodecanoyl)-L-homoserine lactone (*OdDHL*), *N*-(3-hydroxy-hexanoyl)-L-homoserine lactone (3-hydroxy-*HHL*), *N*-(3-hydroxy-dodecanoyl)-L-homoserine lactone (3-hydroxy-*dDHL*)

Included in this list are the human pathogens *Pseudomonas aeruginosa*, *Burkholderia cepacia*, *Chromobacterium violaceum*, *Yersinia pestis*, *Y. entercolitica*, *Y. pseudotuberculosis*, *Aeromonas hydrophila* and *Brucella melitensis*.

The AHL-dependent communication systems share common regulatory features. AHL signal molecules are synthesized from precursors by a synthase protein “I” and interact with transcriptional regulator proteins “R” to regulate expression of target genes. The LuxI-type proteins catalyse the formation of AHLs from the appropriately charged acyl carrier protein as the major acyl chain donor and *S*-adenosyl methionine, which provides the homoserine lactone moiety (Parsek et al. 1999) (Fig. 6). The general theme of QS systems is their role as co-ordinators in the expression of traits that are required for pathogenic and symbiotic interactions with higher organisms. These include expression of bioluminescence in specialized light organs of squids and fish, surface colonization, biofilm development as well as production of virulence factors and hydrolytic enzymes during infection of eukaryotic hosts.

Furanones specifically interfere with AHL-regulated quorum sensing (QS). Givskov et al. (1996) demonstrated that natural furanones inhibit the QS controlled surface motility and colonization of *S. liquefaciens* and hypothesised that this constituted a specific means of eukaryotic interference with bacterial quorum sensing. Extensive experimental evidence in support of this hypothesis has since been presented including furanones repressing AHL-dependent expression of *V. fischeri* bioluminescence (Manefield et al. 1999), inhibiting AHL-controlled virulence factor production and pathogenesis in *P. aeruginosa* (Hentzer et al. 2002, 2003), and QS controlled carbapenem production in *E. carotovora* (Manefield et al. 2001).

Expression of the above phenotypes was inhibited in a concentration-dependent manner at concentrations that did not affect the growth of the test organisms. Furthermore, nuclear magnetic resonance spectrometry was used to confirm that the observed inhibition was not due to a

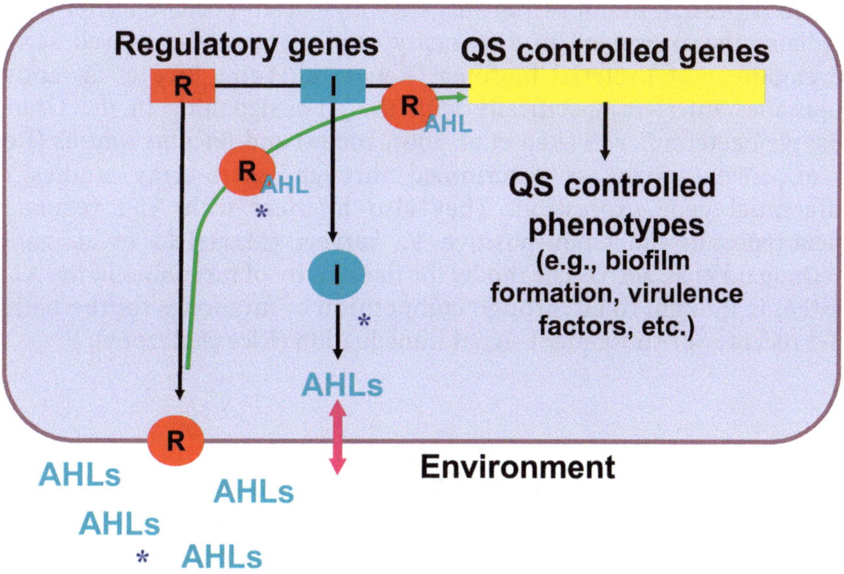

Fig. 6. Schematic of the AHL quorum-sensing pathway. The AHL synthase, shown as a generic "I" gene and protein, produces the AHL enzymatically, which then diffuses in and out of the cell and increases in concentration. Binding of the signal to the membrane-bound receptor, shown as a generic "R" gene and protein, leads to expression of the quorum sensing controlled genes and the quorum sensing controlled phenotypes. Because auto-induction of the "I" gene varies between organisms, it is not included in this model. The asterisk represents potential sites of interference with the AHL system. (Reproduced with permission from Rice et al. 2005)

chemical interaction between AHLs and furanones (Manefield et al. 1999). However, the most direct indications of molecular interactions between furanones and the QS receptor comes from the observations of their ability to displace radio-labelled AHLs from *E. coli* overproducing the LuxR protein (Manefield et al. 1999). Whereas radio-labelled AHL binds to *E. coli* cells overproducing the LuxR protein, a similar experimental design revealed that radio-labelled furanone did not show substantial affinity for cells overproducing LuxR. Western blot analysis using antibodies against LuxR demonstrated that furanones accelerated the degradation of the LuxR receptor in a concentration-dependent manner (Manefield et al. 2002). Expression of a LuxR controlled *PluxI-gfp* reporter correlated with the amount of intact LuxR protein suggesting that furanones inhibit QS by destabilizing the receptor. It is likely that binding of the furanone probably introduces conformational changes in the protein. This in turn leads to recognition by cellular proteases that degrade the mis-folded receptor. This model is supported by the finding that administration of AHL can partly relieve the effect of furanones by protecting the LuxR protein (Manefield et al. 2002).

More recently furanones have also been identified as inhibitors of an alternative signalling system, the autoinducer 2 (AI-2) system. The AI-2 system is present in both Gram-negative and Gram-positive bacteria and regulates the expression of a diversity of virulence traits as well as the development of bacterial biofilms (Ren et al. 2001; Rice et al. 2005). Furanones interfere specifically with the AI-2 signalling in the Gram-positive bacteria *E. coli* (Ren et al. 2001, 2004a) and *Bacillus subtilis* (Ren et al. 2002, 2004b) as determined through micro-array studies of differential gene expression. They also interfere with AI-2 regulated phenotypes in the Gram-positive *V. harveyi* (Manefield et al. 2000; McDougald et al. 2003). The model for the activity of furanones in the AI-2 system is thought to be through competition of furanones for the native AI-2 receptor or subsequent signal transduction (Rice et al. 2005).

4
The Development and Application of Furanones

Biological models such as *Delisea* provide a direct link between biology and biotechnology (de Nys and Steinberg 2002). Furanones have strong and specific *in situ* effects against both micro- and macro-fouling and this has led to the investigation of these compounds as leads for new technologies against both micro- and macro-fouling. These technologies have been developed in parallel streams somewhat akin to the development of pharmaceuticals. However, the understanding of the mechanism of action of furanones in bacterial quorum sensing has meant the research and development of furanones as inhibitors of the development and expression of biofilms has progressed more rapidly than the application of furanones as antifouling technologies against macrofouling. Both streams of research and development rely on the synthesis and supply of furanones. Here we review the chemical synthesis of furanone analogues, their application as agents for the prevention and treatment of bacterial infections, and as a new antifouling technology for marine industries.

4.1 Chemical Synthesis

The aim of the synthesis of furanones has been to develop a library of compounds for biological testing against micro- and macro-fouling and an in-depth investigation of the molecular mechanisms of action of furanones, including QSAR studies, as inhibitors of quorum sensing regulated phenotypes. As a group furanones, or fimbrolides, share a common 4-halo-3-butyl-5-halomethylene-2(5*H*)-furanone skeleton but differ in the number and nature of the halogen substituents and the presence or absence of oxygen functionality in the butyl side chain. There are very few successful syntheses of these novel molecules reported in the literature. The first synthesis of 4-bromo-3-butyl-5-bromomethylene-2(5*H*)-furanone was by Beecham and Sims (1979) and involved the acid-catalysed cyclisation of brominated 2-(2-oxopropyl)hexanoic acid. A reinvestigation of this reaction revealed it not only yields the natural furanone **2** but also significant amounts of isomeric 3-butyl-5-dibromo-methylene-2(5*H*)-furanone **8** (Fig. 7) and tribromo furanone **1** (Manny et al. 1997). This reaction is remarkable because of its success under seemingly harsh conditions of concentrated sulphuric acid at elevated temperature. Caine and Ukachukwu (1985) also developed an elegant synthesis of 4-bromo-3-butyl-5-bromomethylene-2(5*H*)-furanone via the reaction of (*E*)-β-bromo-β-lithioacrylates with acetic anhydride to form 5-hydroxy-5-methyl-4-bromo-3-butyl-2(5*H*)-furanone in high yield which could be further transformed into the desired bromomethylene compound. However, these syntheses provide a limited basis for the development of metabolites as they only addressed the synthesis of the unsubstituted furanones.

Synthesis of more substituted analogues of fimbrolides bearing a hydroxy or an acetoxy functionality at C-1' is more challenging and provides the basis for further substitution and the development of more extensive libraries. This substitution at C-1' has been the focus of research. Oxidation of the suitably substituted furans to the corresponding 5-hydroxy-5-methyl-2(5*H*)-furanones followed by dehydration and bromomethylation has been used successfully for the synthesis of furanones. Jefford et al.

(1989) used 3-bromo-2-methyl-4-(1'-hydroxybutyl)furan, derived from 5-methylfurfural, to synthesise 3-(1'-hydroxybutyl)-4-bromo-5-bromo-methylene-5(2*H*)furanone (Scheme 1), while Kotsuki et al. (1983) prepared a desbromo analogue of 3-(1'-hydroxybutyl)-furanone starting 2-methyl-4-(1'-acetoxybutyl)furan (Scheme 2).

Reaction of γ-monosubstituted allenic ester with *N*-bromosuccinimide has been used by de March et al. (1995) to prepare 4-bromo-3-butyl-5-methyl-2(5*H*)-furanone. Further transformation of this compound by standard conditions followed by the introduction of the bromomethylene group allowed the formation of 3-(1-acetoxybutyl)-4-bromo-5-bromo-methylene-2(5*H*)-furanone in moderate yield (Scheme 3).

The most facile synthesis of substituted fimbrolides was reported by Read and Kumar 1999. The synthesis involves a very efficient free

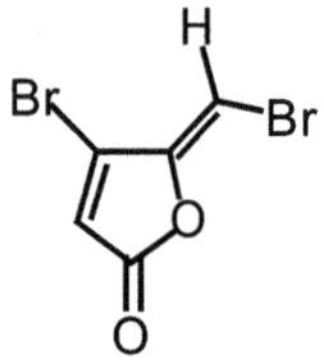

30

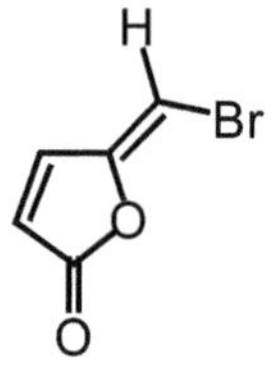

56

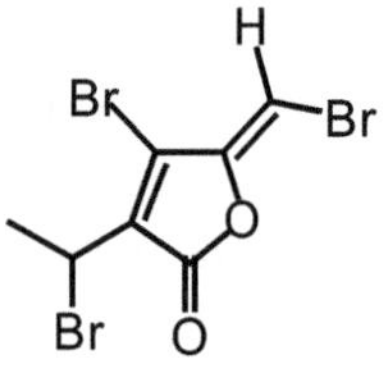

26

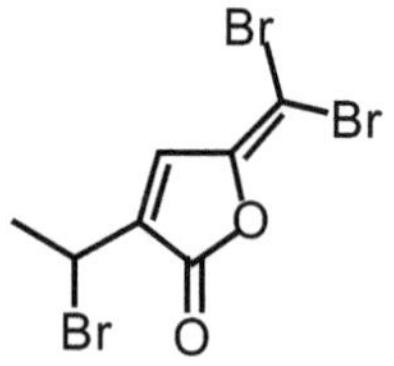

27

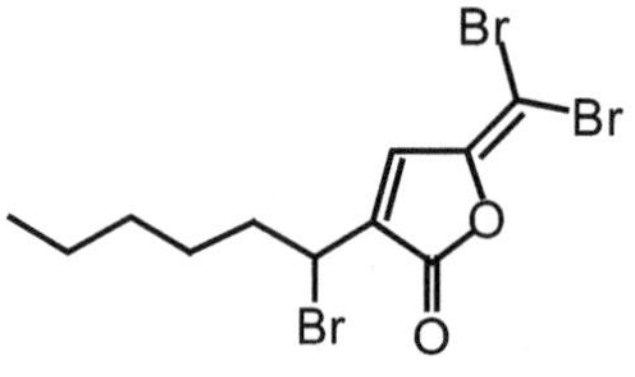

34

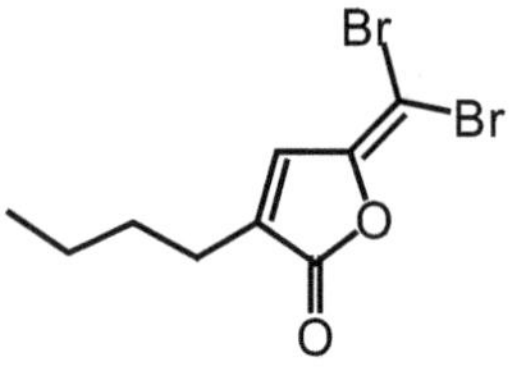

8

Fig. 7. Structures of synthetic furanones

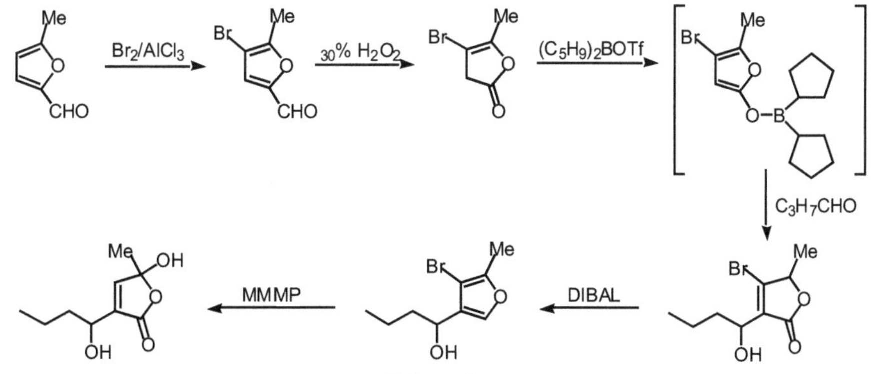

Scheme 1

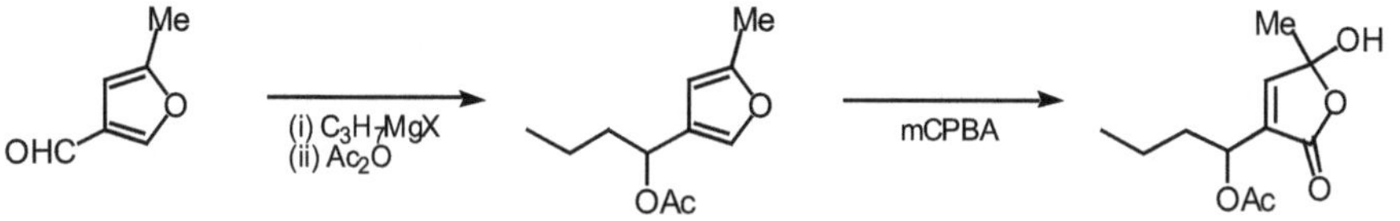

Scheme 2

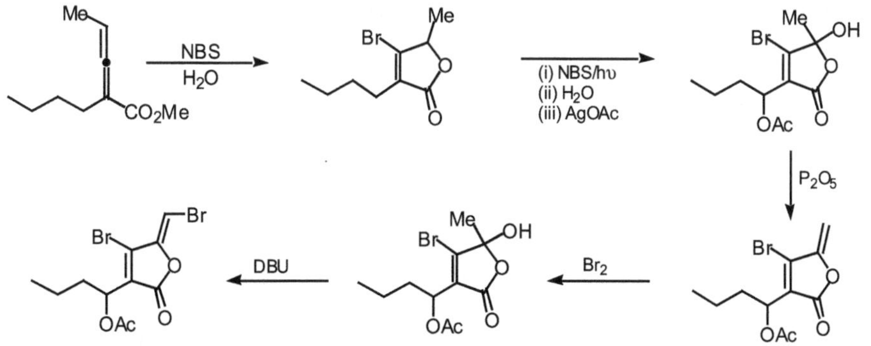

Scheme 3

radical bromination of the furanone **2** with NBS in the presence of catalytic amount of benzoyl peroxide to yield 4-bromo-3-(1'-bromobutyl)-5-bromomethylene-2(5*H*)-furanone in a high yield (Scheme 4). The bromobutyl furanone can then be easily converted into the corresponding hydroxybutyl and acetoxybutyl compound. The hydroxybutyl compound then serves as an ideal substrate for the synthesis of a variety of new analogues. This methodology coupled with our re-investigation of the acid-catalysed cyclisation has allowed the variation of the chemical structure of the furanones in a controlled fashion.

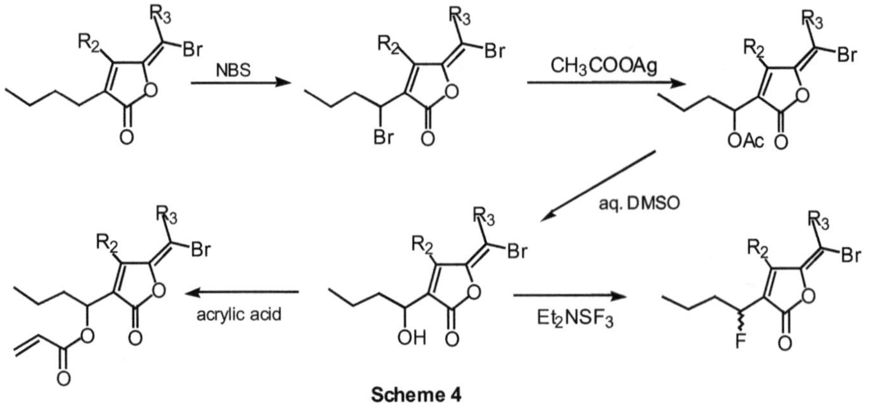

Scheme 4

The aim of these syntheses is to develop the broadest possible range of related compounds for testing against the inhibition of microfouling and biofilm development. In order to develop a comprehensive structure-function hypothesis, a series of furanones where the 3-butyl chain from the C-3 was switched to the C-4 position and the C-4 bromine to the C-3 position were also targeted (Kumar and Read 2002). The synthesis was accomplished via the reaction of 2-alkanone with glyoxalic acid followed by bromination and cyclisation of the 4-oxo-2-alkenoic acid. The resulting 4-alkyl-3,4-dibromo-5-methylenedihydrofuranone was dehydrobrominated and converted to the corresponding 5-bromomethylene furanone by the standard procedure (Scheme 5).

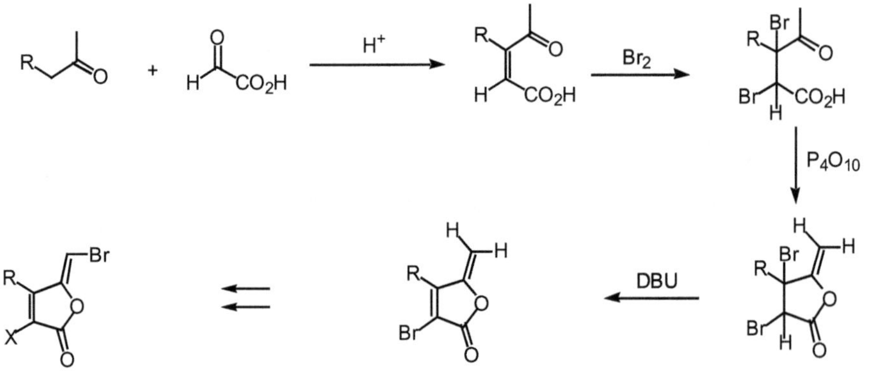

Scheme 5

This has facilitated the development of analogues with variable alkyl side chains at C-3, the electronegative substituent in the alkyl chain, the degree of bromination, and variation of the substitution pattern of the exocyclic 5-bromomethylene group, and oxidation level in the furanone molecule. To date more than 200 analogues of furanones have been synthesised and evaluated for their efficacy in a variety of antimicrobial and biofouling assays.

4.2 Delivery

A recent major goal of the project is to develop techniques for binding of furanones to polymer backbones. This has particular applications in the prevention of biofilms on biomaterials such as contact lenses, catheters and medical implants where covalently bound furanones are advantageous from both a regulatory and commercial perspective (Baveja et al. 2004a,b; Hume et al. 2004). Two methods, co-polymerisation and surface attachment, have been used to incorporate furanones into polymers.

4.2.1 Co-Polymerisation of Furanones

A number of different strategies have been used to covalently attach furanones to polymers. These include atom transfer polymerisation of styrene or acrylate in the presence of a 3-(1'-bromoalkyl)-furanone, and bulk polymerisation of an acrylate in the presence of 3-(1'-acryloyloxyalkyl)-furanone. The polymers prepared by atom transfer methods result in polymers in which the furanones are located at the end of the chain while bulk polymerisation generates co-polymers in which furanones are incorporated into the polymer chain. Recently, co-polymers of styrene containing 0.75%, 2.54%, 4.1% and 6.5% of 3-(1-bromohexyl)-5-dibromomethylene-2(5*H*)-furanone have been synthesised by the atom transfer co-polymerisation process and evaluated for their anti-adhesion properties against *Staphylococcus epidermidis* (Hume et al. 2004). Similarly methyl methacrylate polymers containing up to 3% of furanone have been synthesized and evaluated for their biological efficacy.

4.2.2 Surface Attachment of Furanones

A number of strategies have been investigated for the attachment of furanones onto biomaterial surfaces including Michael addition of surface amino groups to furanone-acrylates, and the nitrene insertion reaction (Read et al. 2001). The latter strategy has been used to attach furanones on to catheters (Hume et al. 2004). The key features of this strategy include plasma activation of the Teflon or catheter surface with an amino group using heptylamine, coupling of the amino groups with poly-acrylic acid, linking 4-azidoaniline via the carboxyl functionality, and finally attachment of the furanone moiety via a nitrene insertion reaction between the furanone and linked azidoaniline (Scheme 6).

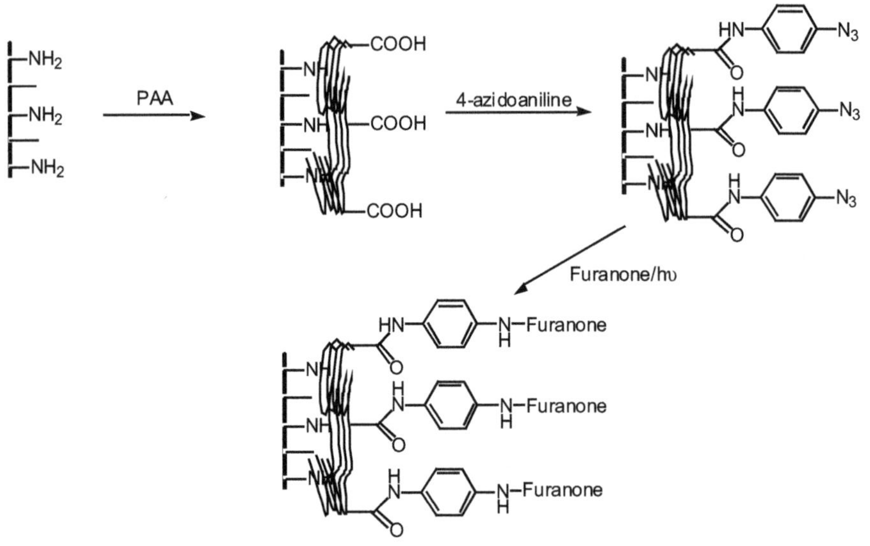

Scheme 6

4.3 Inhibition of Pathogenic Phenotypes and the Development of Anti-Infectives

Quorum-sensing (QS) systems of pathogens are central regulators for the expression of virulence factors and represent highly attractive targets for the development of novel therapeutics. QS systems regulate, in the classical sense, non-essential phenotypes and therefore the inhibition of QS specifically abolishes expression of pathogenic traits but does not affect bacterial growth. This suggests that QS inhibitor (QSI) compounds should not impose the same selective pressure on bacteria as that imposed by antibiotics. However, with production of virulence factors blocked, the bacteria can no longer adapt to the host environment and consequently are cleared by innate host defences.

The idea of using signal molecule-based drugs to control bacterial virulence rather than bacterial growth is attractive for several reasons. The extensive use of antibiotics has lead to an increase in resistant strains, which in turn has limited the effect of the traditional antimicrobial treatment. Furthermore, bacteria predominantly live as surface-associated sessile communities or biofilms (Costerton et al. 1987; Geesey et al. 1997). In clinical microbiology, the biofilm mode of bacterial growth is particularly important as many persistent and chronic infections such as pulmonary infections of cystic fibrosis (CF) patients, periodontitis, otitis media, biliary tract infection, and endocarditis, as well as the colonization of medical implants (catheters, artificial heart valves, etc.)

are intrinsically linked to the formation of bacterial biofilms (Costerton et al. 1999).

With the recent public announcement from the US National Institutes of Health that more than 80% of all microbial infections involve biofilms (Davies 2003) it is necessary to consider the capability of forming a biofilm within the human body to represent a pathogenic trait *per se*. For the clinician, the major problem is that bacteria embedded in the biofilm matrix can often withstand host immune responses and are markedly more tolerant to antibiotics (up to 1000-fold relative to planktonic cultures), often exceeding the highest deliverable doses of antibiotics and thus making an efficient treatment impossible (Costerton et al. 1987, 1999; Høiby et al. 2001; Drenkard 2003).

Furanones have been widely investigated with respect to their impact on pathogenic traits but the greatest focus has been on the Gram-negative pathogen *Pseudomonas aeruginosa*. *P. aeruginosa* is the most common Gram-negative bacterium found in nosocomial and life-threatening infections of immuno-compromised patients (van Delden and Iglewski 1998). Patients with CF are especially disposed to *P. aeruginosa* infections and the bacterium is responsible for high rates of morbidity and mortality (Høiby and Frederiksen 2000; Lyczak et al. 2002). *Pseudomonas aeruginosa* produces a cocktail of virulence factors that contribute to its pathogenesis and QS plays a key role in orchestrating the expression of many of these such as exoproteases, siderophores, exotoxins and several secondary metabolites, and participates in the development of biofilms (Passador et al. 1993; Winson et al. 1995; Davies et al. 1998; Hentzer et al. 2003, 2004). *Pseudomonas aeruginosa* possesses two QS systems: the LasR-LasI and the RhlR-RhlI with the cognate signal molecules OdDHL [*N*-(3-oxo-dodecanoyl)-L-homoserine lactone] and BHL [*N*-(butanoyl)-L-homoserine lactone], respectively. The two QS systems of *P. aeruginosa* are hierarchically arranged with the *las* system being on top of the signalling cascade positively regulating expression of both *rhlR* and *rhlI* (Latifi et al. 1996; Pesci et al. 1997). Natural furanones and their synthetic analogues have been tested for their efficacy as inhibitors of AHL mediated pathogenicity in *P. aeruginosa*. While natural furanones have generated substantial knowledge about structure-function relationships of AHL signals and blockers, they are not potent inhibitors of QS in *P. aeruginosa*. This is because the binding sites of the LasR and RhlR proteins differ and it is therefore difficult to block two QS receptors with one inhibitor (Smith et al. 2003a,b).

Furanone analogues, in particular furanone **30** (Fig. 7), which shows least similarity to the cognate AHL signals, exhibit the most potent QSI activity blocking QS in *P. aeruginosa* in the 1–10 μM range. In contrast to conventional antibiotics, the synthetic furanones are equally active on planktonic as well as biofilm-embedded cells. Without further increase in concentration, expression of an OdDHL-regulated *lasB-gfp* reporter fusion in a characteristic *P. aeruginosa* biofilm could be inhibited by furanones deep into the 'mushroom'-like structures (Hentzer et al. 2002). Another interesting aspect of the furanones is the synergy with conventional antibiotics and detergents. *Pseudomonas aeruginosa* biofilms treated with synthetic furanones such as **30** and **56** (Fig. 7) were also subsequently exposed to SDS or the antibiotic tobramycin (the latter is used to treat infections in CF patients). Pre-treatment with furanones diminished the biofilm tolerance to the detergent or to tobramycin in such a way that these antimicrobials were able to penetrate and kill the majority of *P. aeruginosa* cells in the pre-treated biofilms. In comparison, the effects of SDS and tobramycin on untreated biofilms were minimal (Fig. 8) (Hentzer et al. 2003).

Furanone **30** is highly specific for QS controlled genes based on transcriptomic-based analysis by means of *P. aeruginosa* anti-sense, oligonucleotide microarrays (Affymetrix, Inc.) (Hentzer et al. 2003). Eighty-five genes (1.5%) were repressed and eight genes (0.1%) were activated in response to treatment of *P. aeruginosa* with furanone **30**.

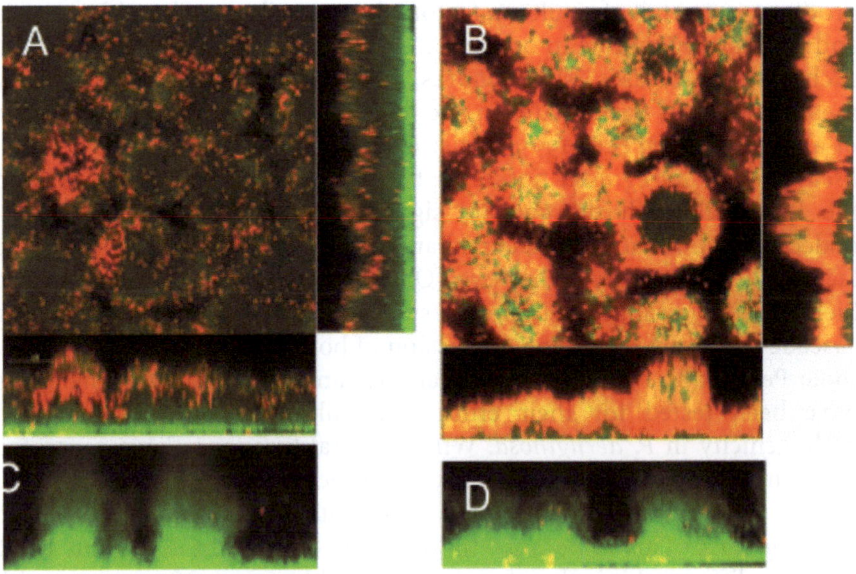

Fig. 8. Sensitivity of furanone **30**-treated *P. aeruginosa* biofilms to tobramycin. Scanning confocal laser microscopy (SCLM) photomicrographs of *P. aeruginosa* PAO1 biofilms grown in the absence (*left panel*) or presence (*right panel*) of furanone **30**. After 3 days, the biofilms were exposed to 100 mg/ml tobramycin for 24 h. Bacterial viability was assayed by staining using the LIVE/DEAD BacLight Bacterial Viability Kit: *red areas* are dead bacteria, and *green areas* are live bacteria. The biofilms were exposed to **A** no furanone and 100 mg ml^{-1} tobramycin, **B** 10 μM furanone **30** and 100 mg ml^{-1} tobramycin, **C** non-treated control and **D** 10 μM furanone **30** and no tobramycin. (reprinted with permission from Hentzer et al. 2003)

Comparative analysis of the furanone **30** target genes and those genes regulated by QS, showed that 85% of the genes that were induced by exogenous OdDHL and BHL were repressed by treatment with **30**. The analysis also demonstrated strong correlation between genes strongly induced and repressed by the two AHLs and **30**, respectively. Among the 85 furanone-repressed genes, 30% have previously been reported as QS-controlled major *P. aeruginosa* virulence factors.

Furanone **30** has also been tested in an *in vivo* mouse pulmonary infection model using live QS monitors, i.e., bacterial cells equipped with a QS regulated target gene fused to a GFP reporter. Mice infected with *P. aeruginosa* and treated with this furanone for 3–5 days following infection showed several orders of magnitude lower bacterial content than a placebo group (Hentzer et al. 2003; Wu et al. 2004). Furthermore, the efficiency of bacterial clearing correlated positively to the concentration of the drug supporting the efficacy and potential of furanone-based treatments for infectious diseases.

4.4 Alternative Modes of Activity

In addition to mediating communication among bacteria the bacterial signalling molecule OdDHL possesses immuno-modulatory properties which can potentially promote a Th-2 dominated response leading to increased tissue damage and inflammation (Telford et al. 1998; Smith et al. 2001). OdDHL also possesses pro-inflammatory, immune modulatory and vaso-relaxant properties (Camara et al. 2002). Recent *in vitro* data indicates that QS blockers also affect the activity of the host defence system (Bjarnsholt et al. 2005). Using an *in vitro* polymorphonuclear neutrophile leucocytes (PMNs)-*P. aeruginosa* biofilm model it was demonstrated that *P. aeruginosa* biofilms resist the activity of the PMNs. However, QS mutants as well as wild-type biofilms treated with furanone **30** were susceptible to PMN phagocytosis (Bjarnsholt et al. 2005; Rasmussen et al. 2005). Activated PMNs liberate oxygen radicals in the form of H_2O_2 which has a bactericidal effect. *P. aeruginosa* biofilm cells tolerate high levels of H_2O_2 (50 to 100 mM) (Elkins et al. 1999; Stewart et al. 2000). Recent data suggests this tolerance results from the amount of oxygen radicals produced by the PMNs, which in turn depends on the magnitude of activation caused by the bacteria. We found that only QS mutants and wild-type biofilms treated with furanone **30** (in contrast to untreated wild-type biofilms) caused detectable activation of the PMNs (Fig. 9) (Bjarnsholt et al. 2005). By combining various QS signal and QSI treatments PMN activation was blocked by the presence of OdDHL and BHL, whereas PMN activation was promoted by the presence of furanone **30** (Bjarnsholt et al. 2005).

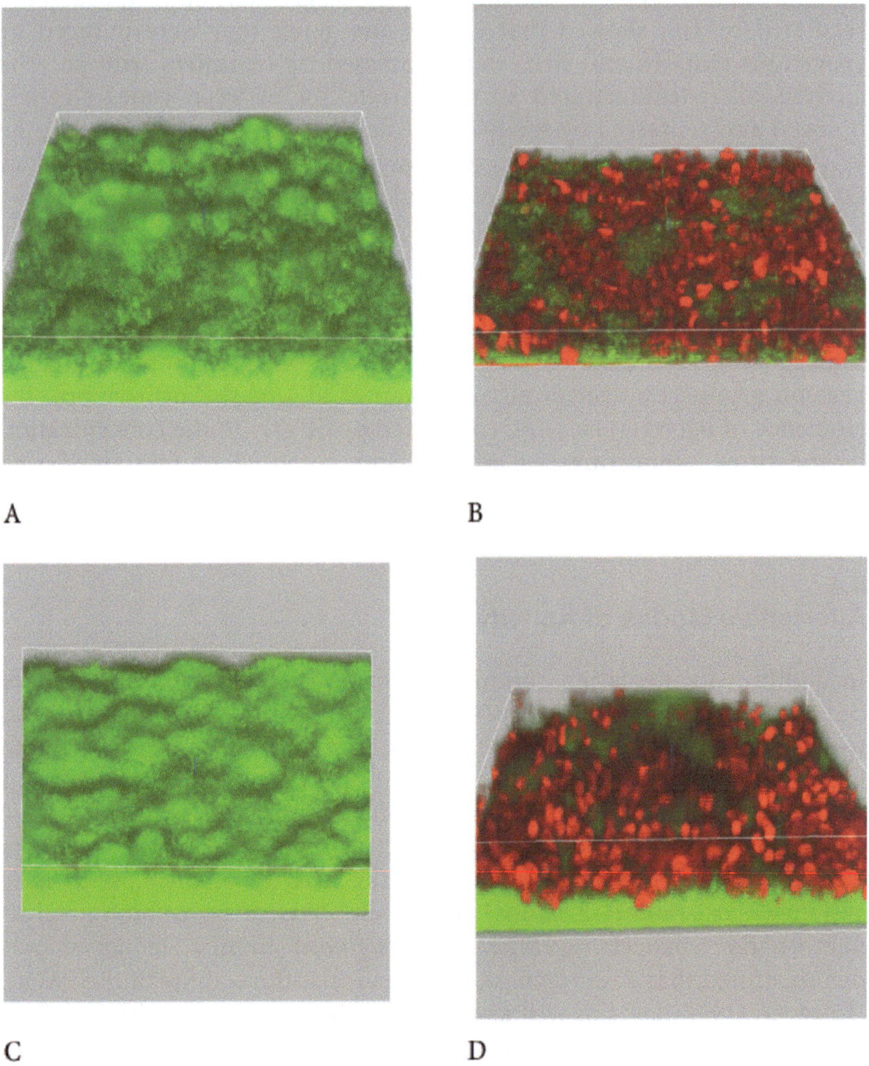

Fig. 9. The QS-dependent tolerance of *P. aeruginosa* PA01 biofilms towards PMN activity. Scanning confocal laser microscopy photomicrographs are shown as 3-D projections. Freshly prepared human PMNs (3×10^6) were introduced into each flow cell channel of 3-day-old biofilms that were exposed to the PMNs for 2.5 h at 37°C. The biofilm cells have green fluorescence due to expression of a GFP tag. The PMNs have red fluorescence through staining with 5 M SYTO 62 (molecular Probes). **A** PA01 biofilm grown in the presence of 10 μM furanone **30** before PMN treatment, **B** after treatment with PMNs. The biofilm is fully penetrated and almost cleared by the PMNs, **C** PA01 biofilm, grown in the absence of furanone treatment, before PMN treatment, **D** after treatment with PMNs. The PMNs reside at the top unable to penetrate the biofilm (Bjarnsholt and Givskov, unpubl.)

This supports a model in which QS interfering signals and AHL signals compete by directly affecting the magnitude of the oxidative PMN burst. This model is also supported by *in vivo* data where pulmonary infections caused by QS mutants induce a faster and stronger immune response against the bacterial infection in the early phase as judged from the severity of lung pathology, higher lung IFN-α production, stronger oxidative burst of blood PMNs, and a faster antibody response compared to the wild-type counterpart (Wu et al. 2001; Bjarnsholt et al. 2005).

In conclusion, a QSI compound such as furanone **30** efficiently eradicates pulmonary infections and also possess PMN activating properties. The administration of QSI compounds can be expected to lead to development of less persistent biofilms but also inhibit expression of bacterial exo-enzymes that actively degrade components of the immune system. Taken together with the synergistic effect of QS blockage and PMN activity, this might promote clearance, which in turn will reverse the severity of infection and improve the lung function. The synergistic effect of antibiotics and QSI drugs may prove to be a useful combination in future chemotherapies. In addition, the ability of furanones to inhibit the development of biofilms may prove effective in medical implants and short-term prophylaxis therapies, while surface coatings containing or liberating QSI compounds might prove efficient in reducing the risk of developing detrimental biofilms on medical devices.

4.5 Biomaterials and Biofilms

The use of quorum sensing inhibitors, such as furanones, in biomaterials offers the potential to prevent the development of detrimental biofilms on biomaterials. This is particular relevance to biomaterials in contact, or implanted into animal tissue and has been demonstrated in principle for selected furanones. The synthetic furanone **34** (Fig. 7) has been coated onto the surface of polymeric biomaterials commonly used in medical devices and tested against the common pathogen *Staphylococcus epidermis* (Baveja et al. 2004a). Furanones physically adsorbed onto biomaterials significantly inhibit the bacterial load on the polymers, in particular on silicon and PTFE, and also reduce slime production with the greatest reduction on silicon (Baveja et al. 2004a). However, there was a consistent leaching or loss of furanones over the 24-h trial period. This has led to the development of covalently bound (non-release) furanones.

Quorum sensing inhibitors that are surface bound, yet still inhibit quorum sensing in pathogenic bacteria, provide for the safest possible use in biomedical applications. Furanones have been incorporated into polymers through co-polymerisation and surface attachment (Sect. 4.2). Furanone **34** when co-polymerised with styrene polymer (0.5 mm diameter) discs (5 and 8.5%) and immobilised on catheter material (Silastic Tenchoff) significantly inhibited biofilm formation of the pathogen *Staphylococcus epidermis* (Hume et al. 2004). Biofilm formation was inhibited by 89% on polystyrene-furanone disks and by 78% on furanone-coated catheters after 24 h, and furanone-coated catheters controlled infection for up to 65 days *in vivo* (Hume et al. 2004). This is in comparison to a commercial antibiotic catheter that inhibited biofilm formation by 26% after 24 h, confirming the potential of furanones in biomaterials. The same hydrolytically stable polystyrene-furanone **34** was also examined for *in vitro* effects on leukocyte (neutrophils and monocyte) cell surface receptor expression and *in vivo* tissue response on neutrophil activation and had a negligible acute inflammatory response either through activation of leukocytes or activation of a tissue response (Baveja et al. 2004b).

The activity of specific furanones, attached directly to the polymer and with long spacer molecules, suggests that they operate through protein receptors that have conserved receptor regions. It is proposed that furanones attached to polymers may bind directly to membrane-bound protein receptors, while furanones linked to polymers via longer spacers inter-collate through the membrane to alternative but conserved receptor proteins. The nature and mechanism of down-regulation of QS regulated phenotypes via interference in conserved receptor domains is an area of continuing research.

4.6 Alternative Quorum Sensing Inhibitor Applications

Given the broad spectrum of applications for quorum sensing inhibitors, furanones have also been investigated for the control of bacterial pathogens in aquaculture (Manefield et al. 2000; Rasch et al. 2004) and in the prevention of corrosion (Ren and Wood 2004). Virulence in aquaculture pathogens, in particular *Vibrio* species, is associated with both AHL and AI-2 quorum sensing systems (Manefield et al. 2000; Rasch et al. 2004; Defroirdt et al. 2004). Furanone **2** decreased the expression of virulence of *V. harveyi* in the black tiger prawn *Penaeus monodon* through the inhibition of toxin production at a concentration (100 μM) that had no effect on growth (Manefield et al. 2000). Similarly furanone **30** (0.001 or 0.1 μM) reduced mortality significantly in the rainbow trout *Oncorhynchus mykiss* when challenged with the pathogen *V. anguillarum* (Rasch et al. 2004). In these trials the growth and survival of *V. anguillarum* was not affected providing for disease control with the likelihood of reduced selection for resistance by the bacteria. Disruption of the quorum sensing mechanism in bacteria has been reviewed and advocated as an alternative strategy for the broader control of aquaculture pathogens (Defoirdt et al.

2004). Furanone **2** has also been investigated for its effect on preventing corrosion by the Gram-positive sulphate reducing bacteria *Desulfotomaculum orientis* (Ren and Wood 2004). Treatment of mild steel incubated with the furanone (40 μg ml^{-1}) reduced weight loss fivefold. While the specific mechanism of action for the inhibition of *D. orientis* remains unknown, furanones interfere in the AI-2 system of Gram-positive bacteria (Rice et al. 2005). Given that Gram-positive bacteria have significant biomedical importance, the effects of furanones on Gram-positive bacteria, particularly through interference with the AI-2 bacterial signalling system offers broad scope for further fundamental and applied research (Ren et al. 2004a,b; Rice et al. 2005).

4.7 Macrofouling – Coatings and Polymers

The most direct link between the biological role of furanones and their application as a marine technology is the action of furanones as antifouling agents against macrofouling. Macrofouling is a well-documented problem for all marine industries ranging from aquaculture of fish and shellfish to shipping and offshore platforms. There is a clear economic driver for environmentally sustainable antifouling solutions across the range of applications. *Delisea* is an obvious model for the development of new antifouling technologies. It is "rarely fouled in the field" and has a lower level of macrofouling than many co-occurring algal species (Nylund et al. 2006). Furanones have a broad spectrum of activity against common cosmopolitan fouling organisms such as the alga *Ulva* sp. and the barnacle *Balanus amphitrite* (de Nys et al. 1995), while surface extracts and furanones inhibit the settlement and growth of a range of epiphytic algal species (Dworjanyn et al. 2006) and bryozoans (Steinberg et al. 2001). Initial laboratory assays determined the activity of furanones as comparable to heavy metals (Cu and Zn) and biocides (Irgarol and Sea-Nine 211) used in commercial antifouling formulations (de Nys et al. 1995). Even with the testing of a limited range of natural furanones in these early studies there is a significant compound-specific activity and a pattern of structure-function activity (de Nys et al. 1995). The lactone moiety of the furanone, as is the case in other natural metabolites with antifouling activity, is required for activity (Rittschof 2001). In conclusion, in laboratory studies natural furanones have a broad range of activity, are in some cases active through a non-toxic mechanism, have compound-specific activities against the suite of fouling organisms tested, have broad variation in chemical properties such as polarity making them amenable to inclusion in a broad range of carriers, and are amenable to synthesis due to their low molecular weight and relative stability.

The move from laboratory assays to more applicable field scale trials is a critical step in the development of natural product-based antifoulants (Rittschof 2001). The first step is normally the further development of synthetic analogues of lead structures to elucidate structure-function relationships, and subsequently identify the most active compounds. This is achieved using synthetic programs combined with laboratory assays based on a moderate throughput. These assays use small amounts of compounds (mg) and provide a quick turnover for the iteration and further development of lead compounds. These tests are often paralleled with safety tests to ensure that leads meet preliminary safety levels concerning cytotoxicity and mutagenicity required for any commercial use chemical. The synthesis program for furanones has provided for a structure-function assessment and the scale-up production of a limited number of lead compounds. This work has also had the advantage of paralleling with safety data from concurrent bacterial and biomedical studies. More than seventy furanones and structurally-related analogues have been assayed against the antifouling test organisms, the algae *Polysiphonia* sp. and *Ulva* sp., the bryozoan *Bugula neretina* and the barnacle *Balanus amphitrite*. Of these compounds, nine have been targeted for further development based on activity and safety.

Throughout this development process, limited subsets of furanones with antifouling activity have been tested in field assays in a range of carriers. The emphasis of most of these studies has been on developing furanones as antifouling agents for aquaculture applications, in particular the prevention of fouling on sea-cage. This is in part because of the desirability of metal free coatings and the need for an effective environmentally acceptable water-based coating.

These experiments have highlighted the key issues of compatibility of compounds with coatings and the "construction" of carriers to regulate release rates (Rittschof 2001). Techniques to control the release rates of furanones include the engineering of the coating using release rate trials, micro-encapsulation (Price et al. 1992) and the co-extrusion of furanones into polymer matrices (Christie et al. 1998). Furanones **30**, and **26** and **27** (as a mixture) (Fig. 7), in solvent-based and solvent-free latex-based coatings on net materials (52-mm mesh) in salmon smoult cages in Norway prevented fouling over a 3-month trial. Controls were heavily fouled by the mussel, *Mytilus edulis*, (452.4±12 mg g^{-1} of net) while the best furanone-based coating **30** inhibited settlement by two orders of magnitude (8.69 mg g^{-1} of net). This was comparable to similar coatings containing Sea-Nine 211 as a positive control (0–4.6 mg g^{-1} of net). There was significant variation between furanones with some being ineffective after three months. Similarly, furanones co-extruded with a range of polymers showed the same variation, with efficacy being dependent on the combination of furanone and polymer (Christie et al. 1998). Furanones were co-extruded using a Haake Rheocode 90 system fitted with a Rheoex TW 100 counterrotation twin-screw extruder fitted with a slit die in 15-cm-wide and 1-mm-thick polymer sheets. The polymer Dupont Elvax 470 (ethylene-vinyl acetate) incorporating 5% of the furanones **26** and **27** (as a mixture) was the most effective of a range of furanone-polymer combinations (Christie et al. 1998) with only 5% cover of the surface area of the polymer after 91 days. This is in comparison to control polymers, which had 100% fouling cover after 28 days (Christie et al. 1998). These trials also established a minimum effective release rate (MERR) over 35 days of 5 μg cm^{-2} day^{-1} for the furanone mixture (**2**, **8** and **1**) to maintain a foul-free surface. This is similar to the MERR for Sea-Nine 211 (5–10 μg cm^{-2} day^{-1}) which provides the benchmark in most applied antifouling studies (Vasishtha et al. 1995). Given the efficacy of furanones in laboratory and field assays, the completion of a broader screening program based on synthetic analogues, baseline data on the compatibility of coatings and polymers on selected furanones, and the breakdown profiles for furanones, there is significant scope to further investigate the application of furanones as an alternative antifouling technology against macrofouling.

4.8 Furanones, Biofouling and Biosignal Ltd.

Furanones are an example of the development of parallel streams of technology to control bacterial- and macro-fouling based on a common platform. The success of research and development of furanones has led to a formalised development and commercialisation program through the University of New South Wales spin-off company Biosignal Ltd. Biosignal Ltd. has full title to 11 patent families and shared title to three patent families protecting key aspects of furanone technology, including the furanones compounds, their manufacturing process and product applications of the compounds.

Acknowledgements. The research programs on furanones have been supported by grants from the Australian Research Council, the National Health and Medical Research Council (Australia), the Department of Environment, Science and Technology (Australia), the Office of Naval Research (USA), the Danish Technical Research Council, the Danish Medical Research Council, the Plasmid Foundation, the Villum Kann-Rasmussen Foundation, the Cystic Fibrosis Foundation, and the German Mukoviszidose e.V.

References

Baveja JK, Willcox MDP, Hume EBH, Kumar N, Odell R, Poole-Warren LA (2004a) Furanones as potential antibacterial coatings on biomaterials. Biomaterials 25: 5003–5012

Baveja JK, Li G, Nordon RE, Hume EBH, Kumar N, Willcox MDP, Poole-Warren LA (2004b) Biological performance of a novel synthetic furanone-based antimicrobial. Biomaterials 25:5013–5021

Beechan CM, Sims JJ (1979) The first synthesis of fimbrolides, a novel class of halogenated lactones naturally occurring in the red seaweed *Delisea fimbriata* (Bonnemaisoniaceae). Tetrahedron Lett 21:1649–1652

Bjarnsholt T, Jensen PO, Burmolle M, Hentzer M, Haagensen JAJ, Hougen HP, Calum H, Madsen KG, Moser C, Molin S, Hoiby N, Givskov M (2005) *Pseudomonas aeruginosa* tolerance to tobramycin, hydrogen peroxide and polymorphonuclear leukocytes is quorum-sensing dependent. Microbiology 151:373–383

Bonin DR, Hawkes MW (1988) Systematics and life histories of New Zealand Bonnemaisoniaceae (Bonnemaisoniales, Rhodophyta). 2. The genus *Delisea*. NZ J Bot 26:619–632

Bouarab K, Kloareg B, Potin P, Correa JA (2001) Ecological and biochemical aspects in algal infectious diseases. Cah Biol Mar 42:91–100

Camara M, Williams P, Hardman A (2002) Controlling infection by tuning in and turning down the volume of bacterial small-talk. Lancet Infect Dis 2:667–676

Caine D, Ukachukwu VC (1985) A new synthesis of 3-n-butyl-4-bromo-5(*Z*)-(bromomethylene)-2-(5*H*)-furanone, a naturally occurring fimbrolide from *Delisea fimbriata* (Bonnemaisoniaceae). J Org Chem 50:2195–2198

Christie GB, Christov V, de Nys R, Steinberg PD, Hodson S (1998) Antifouling polymers. International patent application PCT/AU98/00509, 18 pp

Costerton JW, Cheng KJ, Geesey GG, Ladd TI, Nickel JC, Dasgupta M, Marrie TJ (1987) Bacterial biofilms in nature and disease. Annu Rev Microbiol 41:435–464

Costerton JW, Stewart PS, Greenberg EP (1999) Bacterial biofilms: a common cause of persistent infections. Science 284:1318–1322

Cueto M, Darias J, San-Martin A, Rovirosa J (1997) New acetyl derivatives from Antarctic *Delisea fimbriata*. J Nat Prod 60:279–281

Dahllof I (2002) Molecular community analysis of microbial diversity. Curr Opin Biotechnol 13:213–217

Davies DG, Parsek MR, Pearson JP, Iglewski BH, Costerton JW, Greenberg EP (1998) The involvement of cell-to-cell signals in the development of a bacterial biofilm. Science 280:295–298

Davies D (2003) Understanding biofilm resistance to antibacterial agents. Nat Rev Drug Discov 2:114–122

Defoirdt T, Boon N, Bossier P, Verstraete W (2004) Disruption of bacterial quorum sensing: an unexplored strategy to fight infections in aquaculture. Aquaculture 240:69–88

De March P, Font J, Gracia A, Qingying Z (1995) Easy access to 5-alkyl-4-bromo-2(5*H*)-furanones, an acetoxyfimbrolide, and bromobeckerelide. J Org Chem 60:1814–1822

De Nys R, Steinberg PD (1999) Role of secondary metabolites from algae and seagrasses in biofouling control. In: Fingerman M, Nagabhushanam R, Thompson M-F (eds) Recent advances in marine biotechnology, vol 3. Biofilms, bioadhesion, corrosion and biofouling. Science Publ, Enfield, pp 223–244

De Nys R, Steinberg PD (2002) Linking marine biology and biotechnology. Curr Opin Biotechnol 13:244–248

De Nys R, Wright AD, Konig GM, Sticher O (1993) New halogenated furanones from the marine alga *Delisea pulchra* (cf. *fimbriata*). Tetrahedron 49:11213–11220

De Nys R, Steinberg PD, Willemsen P, Dworjanyn SA, Gabelish CL, King RJ (1995) Broad spectrum effects of secondary metabolites from the red alga *Delisea pulchra* in antifouling assays. Biofouling 8:259–271

De Nys R, Steinberg PD, Rogers CN, Charlton TS, Duncan MW (1996) Quantitative variation of secondary metabolites in the seahare *Aplysia parvula* and its host plant, *Delisea pulchra*. Mar Ecol Prog Ser 130:135–146

De Nys R, Dworjanyn SA, Steinberg PD (1998) A new method for determining surface concentrations of marine products on seaweeds. Mar Ecol Prog Ser 162:79–87

Drenkard E (2003) Antimicrobial resistance of *Pseudomonas aeruginosa* biofilms. Microb Infect 5:1213–1219

Dunny GM, Winans SC (1999) Bacterial life: neither lonely nor boring. In: Dunny GM, Winans SC (eds) Cell-cell signaling in bacteria. Am Soc Microbiol Press, Washington, DC, pp 1–5

Dworjanyn SA (2001) Chemically mediated antifouling and the cost of producing secondary metabolites in a marine alga. PhD Thesis. University of NSW, Sydney

Dworjanyn SA, de Nys R, Steinberg PD (1999) Localisation and surface quantification of secondary metabolites in the red alga *Delisea pulchra*. Mar Biol 133:727–736

Dworjanyn SA, de Nys R, Steinberg PD (2006) Chemically mediated antifouling in the red alga *Delisea pulchra*. Mar Ecol Prog Ser (in press)

Eberl L (1999) *N*-Acyl homoserine lactone-mediated gene regulation in Gram-negative bacteria. Syst Appl Microbiol 22:493–506

Elkins JG, Hassett DJ, Stewart PS, Schweizer HP, McDermott TR (1999) Protective role of catalase in *Pseudomonas aeruginosa* biofilm resistance to hydrogen peroxide. Appl Environ Microbiol 65:4594–4600

Engel S, Jensen PR, Fenical W (2002) Chemical ecology of marine microbial defence. J Chem Ecol 28:1971–1985

Fries L (1966) Influence of iodine and bromine on growth of some red algae in axenic culture. Physiol Plant 19:800–808

Fuqua WC, Winans SC, Greenberg EP (1994) Quorum sensing in bacteria: the LuxR-Luxi family of cell density-responsive transcriptional regulators. J Bacteriol 176:269–275

Geesey GG, Richardson WT, Yeomans HG, Irvin RT, Costerton JW (1997) Microscopic examination of natural sessile bacterial populations from an alpine stream. Can J Microbiol 23:1733–1736

Givskov M, de Nys R, Manefield M, Gram L, Maximilien R, Eberl L, Molin S, Steinberg PD, Kjelleberg S (1996) Eukaryotic interference with homoserine lactone-mediated prokaryotic signaling. J Bacteriol 178:6618–6622

Hentzer M, Riedel K, Rasmussen TB, Heydorn A, Anderson, JB, Parsek MR, Rice SA, Eberl L, Molin S, Hoiby N, Kjelleberg S, Givskov M (2002) Inhibition of quorum sensing in *Pseudomonas aeruginosa* biofilm bacteria by a halogenated furanone compound. Microbiology 148:87–102

Hentzer M, Wu H, Anderson JB, Riedel K, Rasmussen TB, Bagge N, Kumar N, Schembri MA, Song Z, Kristoffersen P, Manefield M, Costerton JW, Molin S, Eberl L, Steinberg P, Kjelleberg S, Hoiby N, Givskov M (2003) Attenuation of *Pseudomonas aeruginosa* virulence by quorum sensing inhibitors. EMBO J 22:3803–3815

Hentzer M, Eberl L, Givskov M (2004) Quorum sensing in biofilms: gossip in the slime world. In: Ghannoum MA, O'Toole G (eds) Microbial biofilms. Am Soc Microbiol Press, Washington, DC, pp 118–141

Høiby N, Frederiksen B (2000) Microbiology of cystic fibrosis. In: Hodson ME, Geddes DM (eds) Cystic fibrosis. Chapman and Hall, London, pp 83–107

Høiby N, Johansen HK, Moser C, Song ZJ, Ciofu O, Kharazmi A (2001) *Pseudomonas aeruginosa* and the *in vitro* and *in vivo* biofilm mode of growth. Microb Infect 3:23–35

Hume EBH, Baveja J, Muir B, Schubert TL, Kumar N, Kjelleberg S, Griesser HJ, Thissen H, Read R, Poole-Warren LA, Schindhelm K, Willcox MDP (2004) The control of *Staphylococcus epidermidis* biofilm formation and *in vivo* infection rates by covalently bound furanones. Biomaterials 25:5023–5030

Jefford CW, Jaggi D, Boukouvalas J (1989) Total synthesis of bromobeckerelide. Tetrahedron Lett 30:1237–1240

Kazlauskas R, Murphy PT, Quinn RJ, Wells RJ (1977) A new class of halogenated lactones from the red alga *Delisea fimbriata* (Bonnemaisoniaceae). Tetrahedron Lett 18:37–40

Kotsuki H, Monden M, Ochi M (1983) Efficient synthesis of acetoxyfimbrolides and beckerilide analogs. Chem Lett 7:1007–1008

Kumar N, Read R (2002) Preparation and antimicrobial activity of fimbrolides. International patent application PCT/AU02/00781, 77 pp

Latifi A, Foglino M, Tanaka K, Williams P, Lazdunski A (1996) A hierarchical quorum-sensing cascade in *Pseudomonas aeruginosa* links the transcriptional activators LasR and RhlR (VsmR) to expression of the stationary-phase sigma factor RpoS. Mol Microbiol 21:1137–1146

Lyczak JB, Cannon CL, Pier GB (2002) Lung infections associated with cystic fibrosis. Clin Microbiol Rev 15:194–222

Manefield M, de Nys R, Kumar N, Read R, Givskov M, Steinberg P, Kjelleberg S (1999) Evidence that halogenated furanones from *Delisea pulchra* inhibit acylated homoserine lactone (AHL)-mediated gene expression by displacing the AHL signal from its receptor protein. Microbiology 145:283–291

Manefield M, Harris L, Rice SA, de Nys R, Kjelleberg S (2000) Inhibition of luminescence and virulence in the black tiger prawn (*Penaeus monodon*) pathogen *Vibrio harveyi* by intercellular signal antagonists. Appl Environ Microbiol 66:2079–2084

Manefield M, Welch M, Givskov M, Salmond GPC, Kjelleberg S (2001) Halogenated furanones from the red alga, *Delisea pulchra*, inhibit carbapenem antibiotic synthesis and exoenzyme virulence factor production in the phytopathogen *Erwinia carotovora*. FEMS Microbiol Lett 205:131–138

Manefield M, Rasmussen TB, Hentzer M, Anderson JB, Steinberg P, Kjelleberg S, Givskov M (2002) Halogenated furanones inhibit quorum sensing through accelerated LuxR turnover. Microbiology 148:1119–1127

Manny AJ, Kjelleberg S, Kumar N, de Nys R, Read RW, Steinberg PD (1997) Reinvestigation of the sulfuric acid-catalysed cyclisation of brominated 2-alkyllevulinic acids to 3-alkyl-5-methylene-2(5*H*)-furanones. Tetrahedron 53:15813–15826

Maximilien R, de Nys R, Holmström C, Gram L, Givskov M, Crass K, Kjelleberg S, Steinberg PD (1998) Chemical mediation of bacterial surface colonisation by secondary metabolites from the red alga *Delisea pulchra*. Aquat Microb Ecol 15:233–246

McCombs JD, Blunt JW, Chambers MV, Munro MHG, Robinson WT (1988) Novel 2(5H)-furanones from the red marine alga *Delisea elegans* (Lamouroux). Tetrahedron 44:1489–1502

McDougald D, Srinivasan S, Rice SA, Kjelleberg S (2003) Signal-mediated cross-talk regulates stress adaptation in *Vibrio* species. Microbiology 149:1923–1933

Millar AJK (1990) Marine red algae of the Coffs harbour region, Northern New South Wales. Aust Syst Bot 6:1–90

Nylund GM, Gribben PE, de Nys R, Pavia H, Steinberg PD (2006) Surface chemistry versus whole cell extracts – the ecological relevance of antifouling tests in seaweeds. Mar Ecol Prog Ser (in press)

Oswald RC, Seed R (1986) Organisation and seasonal progression within the epifaunal communities of coastal macroalga. Cah Biol Mar 27:29–40

Passador L, Cook JM, Gambello MJ, Rust L, Iglewski BH (1993) Expression of *Psuedomonas-aeruginosa* virulence genes requires cell-to-cell communication. Science 260:1127–1130

Parsek MR, Greenberg EP (2000) Acyl-homoserine lactone quorum sensing in Gram-negative bacteria: a signalling mechanism involved in associations with higher organisms. Proc Natl Acad Sci USA 97:8789–8793

Parsek MR, Val DL, Hanzelka BL, Cronan JE, Greenberg EP (1999) Acyl homoserine-lactone quorum-sensing signal generation. Proc Natl Acad Sci USA 96:4360–4365

Pesci EC, Pearson JP, Seed PC, Iglewski BH (1997) Regulation of *las* and *rhl* quorum sensing in *Pseudomonas aeruginosa*. J Bacteriol 179:3127–3132

Pettus JA, Wing RM, Sims JJ (1977) Marine natural products XII Isolation of a family of multihalogenated gamma-methylene lactones from the red seaweed *Delisea fimbriata*. Tetrahedron Lett 18:41–44

Potin P, Bouarab K, Salaun JP, Pohnert G, Kloareg B (2002) Biotic interactions of marine algae. Curr Opin Plant Biol 5:308–317

Price RR, Patchan M, Clare A, Rittschof D, Bonadventure J (1992) Performance enhancement of natural antifouling compounds and their analogs through microencapsulation and controlled release. Biofouling 6:207–216

Rasch M, Buch C, Austin B, Slierendrecht WJ, Ekmann KS, Laresen JL, Johansen C, Riedle K, Eberl L, Givskov M, Gram L (2004) An inhibitor of bacterial quorum sensing reduces mortalities caused by vibriosis in rainbow trout (*Oncorhynchus mykiss*, Walbaum). Syst Appl Microbiol 27:350–359

Rasmussen T, Skindersø ME, Bjarnsholt T, Phipps RK, Christensen KB, Andersen JB, Ostenfeld-Larsen T, Hentzer M, Givskov M (2005) Identity and effects of quorum sensing inhibitors produced by *Penicillum* species. Microbiology 151 (in press)

Read R, Kumar N (1999) Preparation of fimbrolide analogues fouling inhibitors and bactericides. International patent application PCT/AU99/00285, 48 pp

Read R, Kumar N, Willcox M, Zhu H, Griesser H, Muir B, Thissen H, Hughes T (2001) Antimicrobial Coatings. International patent application PCT /AU01/0047, 23 pp

Ren D, Wood TK (2004) (5*Z*)-4-bromo-5-(bromomethylene)-3-butyl-2(5*H*)-furanone reduces corrosion from *Desulfotomaculum orientis*. Environ Microbiol 6:535–540

Ren D, Sims JJ, Wood TK (2001) Inhibition of biofilm formation and swarming of *Escherichia coli* by (5*Z*)-4-bromo-5-(bromomethylene)-3-butyl-2(5*H*)-furanone. Environ Microbiol 3:731–736

Ren D, Sims JJ, Wood TK (2002) Inhibition of biofilm formation and swarming of *Bacillus subtilis* by (5*Z*)-4-bromo-5-(bromomethylene)-3-butyl-2(5*H*)-furanone. Lett Appl Microbiol 34:293–299

Ren D, Bedzyk LA, Ye RW, Thomas SM, Wood TK (2004a) Differential gene expression shows natural brominated furanones interfere with the auto-inducer-2 bacterial signalling system of *Escherichia coli*. Biotechnol Bioeng 88:630–642

Ren D, Bedzyk LA, Setlow P, England DF, Kjelleberg S, Thomas SM, Ye RW, Wood TK (2004b) Differential gene expression to investigate the effect of (5*Z*)-4-bromo-5-(bromomethylene)-3-butyl-2(5*H*)-furanone on *Bacillus subtilis*. Appl Environ Microbiol 70:4941–4949

Rice SA, McDougald D, Kumar N, Kjelleberg S (2005) The use of quorum sensing blockers as therapeutic agents for the control of biofilm associated infections. Curr Opin Invest Drugs 6:178–184

Rittschof D (2001) Natural products antifoulants and coatings development. In: McClintock JB, Daker BJ (eds) Marine chemical ecology. CRC Press, Boca Raton, pp 543–566

Rogers CN, de Nys R, Charlton TS, Steinberg PD (2000) Dynamics of algal secondary metabolites in two species of sea hare. J Chem Ecol 26:721–74

Smith RS, Fedyk ER, Springer TA, Mukaida N, Iglewski BH, Phipps RP (2001) IL-8 production in human lung fibroblasts and epithelial cells activated by the *Pseudomonas* autoinducer *N*-3-oxododecanoyl homoserine lactone is transcriptionally regulated by NF-kappa B and activator protein-2. J Immunol 167:366–374

Smith KM, Bu YG, Suga H (2003a) Induction and inhibition of *Pseudomonas aeruginosa* quorum sensing by synthetic autoinducer analogs. Chem Biol 10:81–89

Smith KM, Bu YG Suga H (2003b) Library screening for synthetic agonists and antagonists of a *Pseudomonas aeruginosa* autoinducer. Chem Biol 10:563–571

Steinberg PD, de Nys R, Kjelleberg S (2001) Chemical mediation of surface colonisation. In: McClintock JB, Baker BJ (eds) Marine chemical ecology. CRC Press, Boca Raton, pp 355–387

Stewart PS, Roe F, Rayner J, Elkins JG, Lewandowski Z, Ochsner UA, Hassett DJ (2000) Effect of catalase on hydrogen peroxide penetration into *Pseudomonas aeruginosa* biofilms. Appl Environ Microbiol 66:836–838

Swift S, Throup JP, Williams P, Salmond GPC, Stewart GSAB (1996) Quorum sensing: a population-density component in the determination of bacterial phenotype. Trends Biochem Sci 21:214–219

Telford G, Wheeler D, Williams P, Tomkins PT, Appleby P, Sewell H, Stewart GS, Bycroft BW, Pritchard DI (1998) The *Pseudomonas aeruginosa* quorum-sensing signal molecule *N*-(3-oxododecanoyl)-L-homoserine lactone has immunomodulatory activity. Infect Immun 66:36–42

Van Delden C, Iglewski BH (1998) Cell-to-cell signaling and *Pseudomonas aeruginosa* infections. Emerg Infect Dis 4:551–560

Vasishtha N, Sundberg DC, Rittschoff D (1995) Evaluation of release rates and control of biofouling using monolithic coatings containing an isothiazolone. Biofouling 9:1–16

Wahl M, Hay ME (1995) Associational resistance and shared doom: effects of epibiosis on herbivory. Oecologia 102:329–340

Whitehead K, Karentz D, Hedges JI (2001) Mycosporine-like amino acids (MAAs) in phytoplankton, a herbivourous pteropod (*Limacina helicinae*), and its pteropod predator (*Clino antarctica)* in McMurdo Bay, Antartica. Mar Biol 139:1013–1019

Williamson JE, Carson DG, de Nys R, Steinberg PD (2004) Demographic consequences of an ontogenetic shift by a sea urchin in response to host plant chemistry. Ecology 85:1355–1371

Winson MK, Camara M, Latifi A, Foglino M, Chhabra SR, Daykin M, Bally M, Chapon V, Salmond GPC Bycroft BW, Lazdunski A, Stewart GSAB, Williams P (1995) Multiple *N*-Acyl-l-homoserine lactone signal molecules regulate production of virulence determinants and secondary metabolites in *Psuedomonas aeruginosa.* Proc Natl Acad Sci USA 92:9427–9431

Wright JT, de Nys R, Steinberg PD (2000) Geographic variation in halogenated furanones from the red alga *Delisea pulchra* and associated herbivores and epiphytes. Mar Ecol Prog Ser 207:227–24

Wright JT, de Nys R, Poore AGB, Steinberg PD (2004) Chemical defence in a marine alga: heritability and the potential for selection by herbivores. Ecology 85:2946–2959

Wu H, Song Z, Givskov M, Doring G, Worlitzsch D, Mathee K, Rygaard J, Høiby N (2001) *Pseudomonas aeruginosa* mutations in *lasI* and *rhlI* quorum sensing systems result in milder chronic lung infection. Microbiology 147:1105–1113

Wu H, Song Z, Hentzer M, Anderson JB, Molin S, Givskov M, Høiby N (2004) Synthetic furanones inhibit quorum-sensing and enhance bacterial clearance in *Pseudomonas aeruginosa* lung infection in mice. J Antimicrob Chemother 53:1054–1061

Isocyano Compounds as Non-Toxic Antifoulants

Y. Nogata, Y. Kitano

Abstract. The marine sponge *Acanthella cavernosa* and nudibranchs of the family Phyllidiidae contain isocyanoterpenoids and their related compounds that show potent antifouling activity against cypris larvae of the barnacle *Balanus amphitrite*, while their toxicity to cyprids is weak. To develop non-toxic antifoulants based on isocyanoterpenoids, especially 3-isocyanotheonellin, a total of 36 isocyano compounds have been synthesized. They were evaluated by both antifouling activity and toxicity toward *B. amphitrite* cyprids, which led some insight into the structure-activity relationships. Since linear alkyl isocyanides showed antifouling activity at nontoxic concentrations, a large amount of 1,1-dimethyl-10-undecyl isocyanide was synthesized, incorporated into paints, and tested for antifouling activity in the field with promising results. Therefore, isocyano compounds were considered as candidate non-toxic antifouling agents.

1 Introduction

Marine sponges have received much attention by natural products chemists and biologists, because they are a rich source of secondary metabolites with novel structures and potent biological activities (Harper et al. 2001; Faulkner 2002; Blunt et al. 2004). The reason why many sponges produce a wide range of secondary metabolites is assumed to be due to their chemical defense system (Bakus et al. 1986; Pawlik 1993; Abarzua and Jakubowski 1995; Engel and Pawlik 2000). Therefore, marine sponges are thought to be ideal targets of discovering potential antifouling agents. In fact, a variety of natural products with antifouling activity have been isolated from marine sponges (Clare 1996; Rittschof 2001; Omae 2003; Fusetani 2004). Particularly interesting as potential

Y. Nogata
Environmental Science Research Laboratory, Central Research Institute of Electric Power Industry, 1646 Abiko, Abiko-shi, Chiba-ken 270-1194, Japan

Y. Kitano
Laboratory of Bio-organic Chemistry, Tokyo University of Agriculture and Technology, 3-5-8 Saiwai-cho, Fuchu, Tokyo 183-8509, Japan

Progress in Molecular and Subcellular Biology
Subseries Marine Molecular Biotechnology
N. Fusetani, A.S. Clare (Eds.): Antifouling Compounds

antifoulants are isocyanoterpenes and related compounds, such as isothiocyanates and formamides among antifouling compounds, which inhibit barnacle settlement at very low concentrations without toxicity.

This chapter describes antifouling isocyanoterpenoid and related compounds isolated from the marine sponge *Acanthella cavernosa* and nudibranchs belonging to the family Phyllidiidae and the structure-activity relationships of simple isocyano compounds obtained by chemical synthesis.

2
Natural Marine Isocyanides

Compounds containing an isocyano group, which are a rare class of natural products, have been isolated from certain marine sponges of the families Axinellidae and Harichondriidae and nudibranches that feed on these sponges (Fusetani 2004; Fusetani et al. 1996; Garson and Simpson 2004). Since the discovery of terpenoid isocyanides from marine sponges in 1973, nearly 200 compounds containing isocyano and related functionalities have been reported. Most of these compounds are toxic toward fish and crustaceans, suggesting a role in chemical defense. In addition, they showed a variety of biological activities including antimicrobial, cytotoxic, and antimalarial (Fusetani et al. 1996; Garson and Simpson 2004). Therefore, it appears reasonable to assume that isocyanides are important to the chemical defense system and perhaps play a role as repellants of epibiotic organisms. Before Fusetani and co-workers (Okino et al. 1995) showed antifouling activity against cypris larvae of the barnacle *Balanus amphitrite*, only Thompson et al. (1985) had described the antifouling activity of isocyano sesquiterpene mixtures derived from an *Axinella* sponge against larvae of a bryozoan, a polychaete, and an abalone.

3
Natural Antifouling Isocyanoterpenes

3.1
Sesquiterpenes

As shown in Fig. 1 and Table 1, a number of isocyano and related sesquiterpenes which showed antifouling activity were isolated from the marine sponge *Acanthella. cavernosa* and four nudibranchs, *Phyllidia pustulosa*, *P. ocelata*, *P. varicose*, and *Phillidiopsis krempfi* (Fusetani et al. 1996; Okino et al. 1996a). Among these sesquiterpenes, 3-isocyanotheonellin (1), which inhibited settlement of cypris larvae with an EC_{50} values of 0.13 µg/mL without significant toxicity (LD_{50} >100 µg/mL), was regarded as a promising antifouling compound (Okino et al. 1996a). 10- Isocyano-4-cadinene (2) (Okino et al. 1996a) and (1*S*, 4*S*, 7*R*, 10*S*)-10-isocyano-5-cadinene-4-ol (3) (Hirota et al. 1998) also showed potent antifouling activity against *B. amphitrite* cyprids with EC_{50} values of 0.14 and 0.17 µg/mL, respectively, whereas their toxicity toward cyprids was very weak (LD_{50} >100 µg/mL). 2-Isocyanotrachyopsane (4) (Okino et al. 1996a), 10-isocyano-4-amorphene (5) (Okino et al. 1996a), and 10-formamido-4-cadinene (6) (Nogata et al. 2003) inhibited larval settlement with EC_{50} values ranging from 0.33 to 0.70 µg/mL. Other isocyanosesquiterpenes were slightly less active than the above compounds (Table 1).

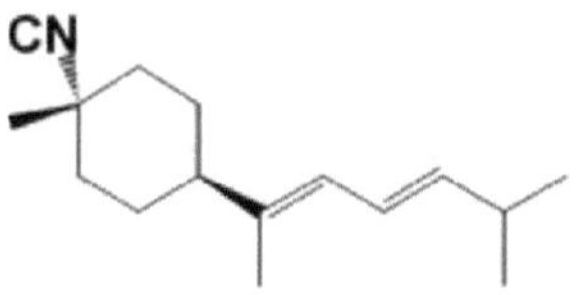

1

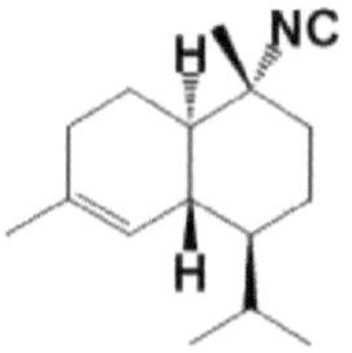

2

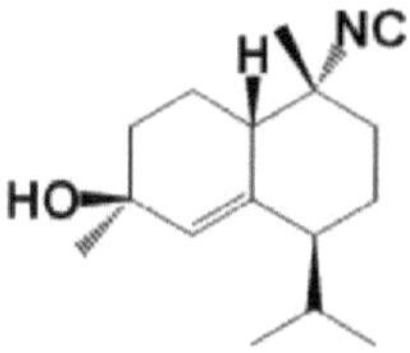

3

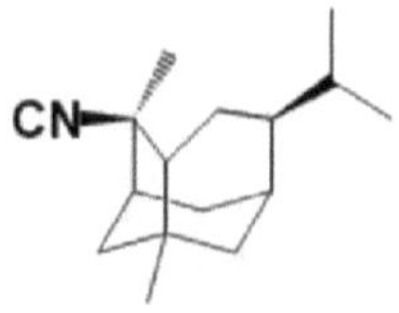

4

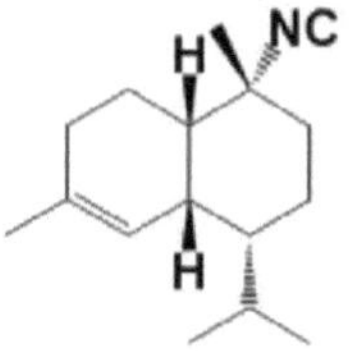

5

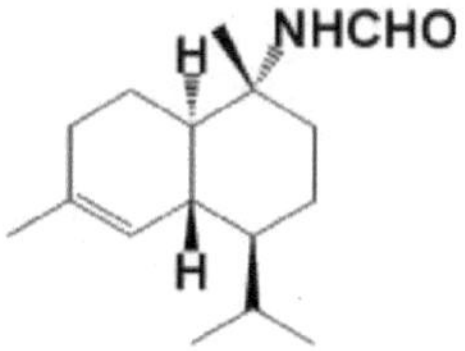

6

Fig. 1. Structures of antifouling sesquiterpenes

Table 1. Antifouling activity and toxicity of sesquiterpenes against cypris larvae

Compound No.	EC_{50} (µg/mL)	LD_{50} (µg/mL)	Reference
3-Isocyanotheonellin (1)	0.13	>100	Fusetani et al. (1996)
10-Isocyano-4-cadinene (2)	0.14	>100	Fusetani et al. (1996)
(1*S*,4*S*,7*R*,10*S*)-10-isocyano-5-cadinene-4-ol (3)	0.17	n.d.	Hirota et al. (1998)
2-Isocyanotrachyopane (4)	0.33	n.d.	Fusetani et al. (1996)
10-Isocyano-4-amorphane (5)	0.70	n.d.	Fusetani et al. (1996)
10-Formamido-4-cadinene (6)	0.50	5.0	Nogata et al. (2003)
(-)-10-Isothiocyanate-4-amorphene	7.2	n.d.	Fusetani et al. (1996)
Axisonitrile-3	3.2	n.d.	Fusetani et al. (1996)
10-*epi*-Axisonitrile	10	n.d.	Fusetani et al. (1996)

3.2 Diterpenes

Thirteen kalihinane-type diterpenes, e.g., kalihinene X(**7**), kalihinene Y (**8**), kalihinene Z (**9**), 15-formamidokalihinene (**10**), 10-formamidokalihinene (**11**), kalihinol A (**12**), kalihinol E (**13**), 10β-formamido-kalihinol-A (**14**), 10β-formamido-kalihinol-E (**15**), 10β-formamido-5β-isocyanatokalihinol-A (**16**), 10β-formamido-5β-isothiocyanatokalihinol-A (**17**), kalihipyran A (**18**), and kalihipyran B (**19**) (Fig. 2), were obtained as antifouling compounds from the marine sponge *A. cavernosa* (Okino et al. 1995, 1996b; Fusetani et al. 1996; Hirota et al. 1996). The antifouling activity of these compounds is summarized in Table 2. Among these compounds, **11–13**, and **17** showed potent antifouling activity (EC_{50} <0.1 μg/mL); more effective and with lower toxicity than copper sulfate. The formamido-substituted kalihinols **14** and **15** were also active with EC_{50} values of 0.12 μg/mL. Kalihipyrans **18** and **19** showed moderate activity with EC_{50}'s of 1.3 and 0.85 μg/mL, respectively.

4 The Structure-Activity Relationships of Synthetic Isocyano Compounds

Yields of the natural antifouling isocyanoterpenoids derived from marine sponges (described above) were generally poor, which hampered their further development as antifouling agents. Moreover, candidate natural products generally have complicated structures, which make it difficult to

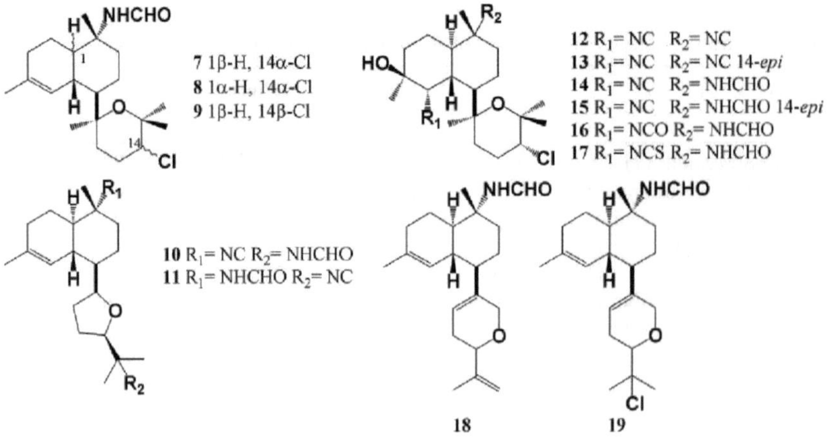

Fig. 2. Structures of antifouling diterpenes

Table 2. Antifouling activity and toxicity of kalihinene- and kalihinol-type diterpenes against cypris larvae

Compound No.	EC_{50} (µg/mL)	LD_{50} (µg/mL)	reference
Kalihinene X (**7**)	0.49	n.d.	Fusetani et al. (1996)
Kalihinene Y (**8**)	0.45	n.d.	Fusetani et al. (1996)
Kalihinene Z(**9**)	1.1	n.d.	Fusetani et al. (1996)
15-Formamidokalihinene (**10**)	0.14	>100	Fusetani et al. (1996)
10-Formamidokalihinene (**11**)	0.095	>100	Fusetani et al. (1996)
Kalihinol A (**12**)	0.087	>100	Fusetani et al. (1996)
Kalihinol E (**13**)	0.088	n.d.	Fusetani et al. (1996)
10β-Formamido-kalihinol-A (**14**)	0.12	n.d.	Fusetani et al. (1996)
10β-Formamido-kalihinol-E (**15**)	0.12	n.d.	Fusetani et al. (1996)
10β-Formamido-5β-isocyanatokalihinol-A (**16**)	0.74	n.d.	Fusetani et al. (1996)
10β-Formamido-5β-isothiocyanatokalihinol-A (**17**)	0.088	n.d.	Fusetani et al. (1996)
Kalihipyran A (**18**)	1.3	n.d.	Fusetani et al. (1996)
Kalihipyran B (**19**)	0.85	n.d.	Fusetani et al. (1996)

obtain sufficient quantities by chemical synthesis. Possible ways to overcome this issue are cell culture and aquaculture of antifoulant-producing organisms as well as analogue development from lead compounds (Clare et al. 1999; Fusetani 2004). As for compounds with simple structures, analogue development is considered as a possible solution to provide adequate quantities for application as antifoulants.

On the other hand, the role of the isocyano group in inhibiting settlement of barnacle larvae is not clear, though many isocyanides have been reported. Studies of the structure-activity relationships seem to be one of the most useful methods to solve this issue.

In order to obtain efficacious, environmentally benign antifouling isocyanides and to investigate the role of the isocyano group in antifouling activity, attempts have been made to synthesize a number of simple isocyano compounds and to evaluate their settlement inhibition activity against *B. amphitrite* cyprids.

4.1 Antifouling Activities of 3-Isocyanotheonellin and Analogues

3-Isocyanotheonellin (**1**), which is a sesquiterpene of the bisabolene class having an isocyano group at the C-3 position, exhibits potent antifouling activity, despite its simple structure (Fusetani et al. 1996; Okino et al. 1996a; Fusetani 2004). This simple structure is suitable for further studies on the structure-activity relationships, and its analogues can be synthesized easily.

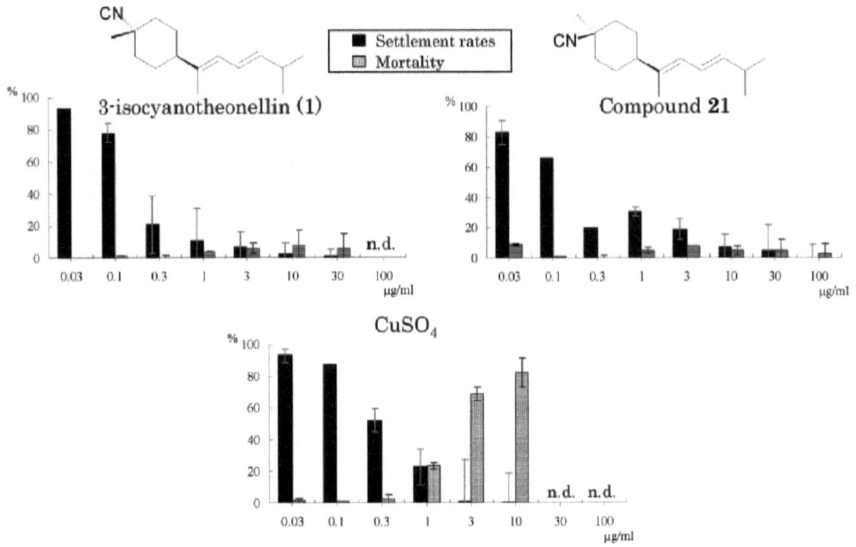

Fig. 3. Effects of synthesized 3-isocyanotheonellin (**1**), **21**, and $CuSO_4$ on settlement and mortality of cypris larvae after 120 h exposure. Rate of settlement of cyprids and mortality in different concentrations are shown as means ± SD of 3–5 replicates

For a structure-activity relationship study, **1** and its analogues, which are the stereoisomer and regioisomer, were initially prepared by the same synthetic route, and their antifouling activity was evaluated against *B. amphitrite* cyprids (Kitano et al. 2002, 2003). Their structures and antifouling activity are illustrated in Table 3, along with the antifouling activity of $CuSO_4$ as a control. Compound **21**, a stereoisomer of **1**, was as active as **1** (Fig. 3). The antifouling activities of **1** and **21** were higher than that of $CuSO_4$. (*Z,E*)-dienes, **20** and **22**, the geometrical isomers of **1** and **21**, respectively, also exhibited potent activity, although they were slightly weaker than those of **1** and **21**. It should be noted that all compounds showed low mortality rates at high concentrations, whereas the LD_{50} of $CuSO_4$ was 2.95 µg/mL (Fig. 3). These results suggest that the stereochemistry of isocyano group at C-3 and the geometry of the butadiene portion do not affect the activity and toxicity.

To clarify the role of the butadiene moiety in the antifouling activity, the synthesis of compounds lacking the methyl group at the C-14 position **23**, as well as its reduced compounds, **24**, **25**, and **26**, were carried out (Kitano et al. 2003). Their structures and antifouling activity are displayed in Table 3. Compounds **23–26** exhibited moderate antifouling activity with EC_{50} values of 1.80, 7.20, 1.80, and 3.80 µg/mL respectively and caused low mortality at high concentrations. Compound **23**, lacking the C-14 methyl group in **21**, was 10-fold less active than **21**, suggesting that the methyl group at the C-14 position influenced the activity. Although the reason is unclear,

Table 3. Structures of 3-isocyanotheonellin (1), its analogues **20-26** and their antifouling activity against cyprids of *B. amphitrite* after 120-h exposure (EC_{50} and LD_{50}). The antifouling activity of isocyanides and $CuSO_4$ is expressed as an EC_{50} value, which defines the concentration that reduces the larval settlement to 50% of the control. The toxicity of compounds is expressed as an LC_{50} value, which defines the concentration that results in 50% mortality.

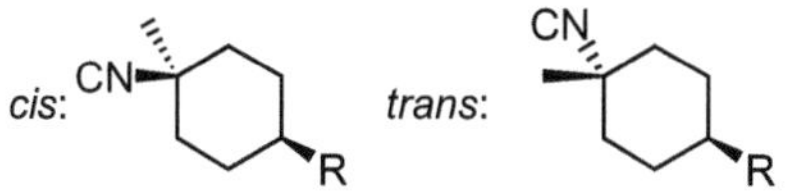

Compound No.	Geometry	R	EC_{50} (μg/ml)	LD_{50} (μg/ml)
1	*trans*	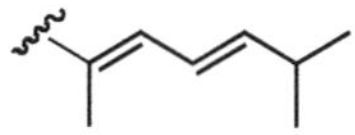	0.19	>100
20	*trans*	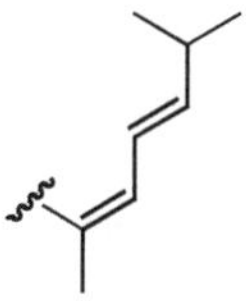	0.29	>100
21	*cis*	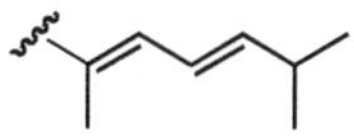	0.18	>100
22	*cis*	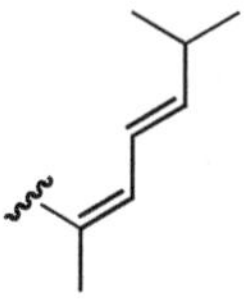	0.41	>100
23	*cis*	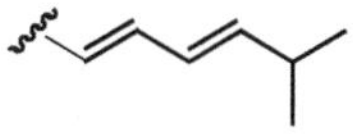	1.80	>100
24	*cis*	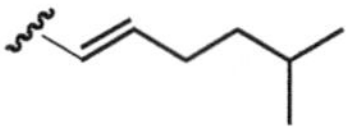	7.20	>100
25	*cis*	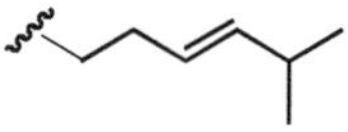	1.80	>100
26	*cis*	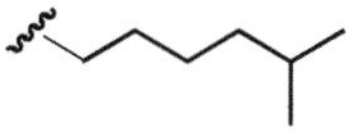	3.80	>100
$CuSO_4$	-	-	0.30	2.95

the methyl group may affect the conformation of the molecule. On the other hand, the double bonds in the side chain did not influence the activity.

4.2
Isocyanocyclohexanes

Isocyanocyclohexanes, which possess an oxygenic functional group at the C-4 position in place of the isoprenyl side chain, were then synthesized and evaluated for their antifouling activity (Kitano et al. 2004). Their structures, antifouling activity and toxicity are summarized in Table 4. Alcohols **27** and **28** and esters **29–34** showed potent antifouling activity without significant toxicity. In particular, most of the synthesized esters exhibited extremely potent antifouling activity; the EC_{50} values were lower than that of $CuSO_4$. It should be noted that the acetate **30** and pivaloate **34** were about 10 times more active than **1** and **21** (Fig. 4); the acetate **30** showed the most potent among synthesized compounds (EC_{50} = 0.0096 μg/mL). On the other hand, ethers **35** and **36** showed moderate activity. The EC_{50} values of the neopentyl ethers **35** and **36** were about 10-fold higher than those of the pivaloates **33** and **34**, although the structures of pivaloates and neopentyl ethers are similar. These results suggest that ester groups at the C-4 position of the isocyanocyclohexane and its stereochemistry are important for antifouling activity. Interestingly, the stereochemistry of alkyl esters exhibited the opposite results of the phenyl esters.

The role of the isocyano group was clearly revealed by the activity of the synthetic des-isocyano derivatives **37** and **38**, where the isocyano group in pivaloates **33** and **34** was missing as shown in Table 4.

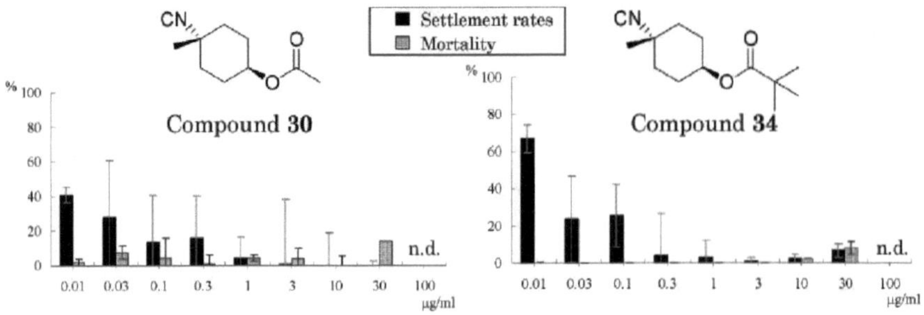

Fig. 4. Effects of synthesized isocyanocyclohexane **30** and **34** on settlement and mortality of cypris larvae after 120 h exposure. Rate of settlement of cyprids and mortality in different concentrations are shown as means ± SD of 3–5 replicates

Table 4. Structures of isocyanocyclohexane **27-36** and des-isocyano pivaloates **37** and **38** and their antifouling activity against cyprids of *B. amphitrite* after 120-h exposure (EC_{50} and LD_{50}). The antifouling activity of compounds is expressed as an EC_{50} value, which defines the concentration that reduces the larval settlement to 50% of the control. The toxicity of compounds is expressed as an LC_{50} value, which defines the concentration that results in 50% mortality.

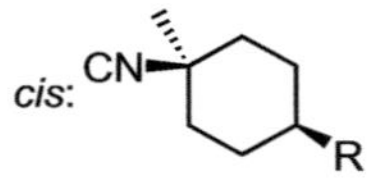

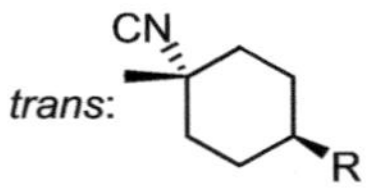

R	Compound No.	Geometry	EC_{50} (μg/ml)	LD_{50} (μg/ml)	Compound No.	Geometry	EC_{50} (μg/ml)	LD_{50} (μg/ml)
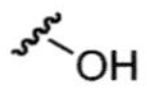	**27**	*cis*	0.98	>30	**28**	*trans*	0.48	>100
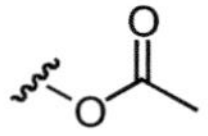	**29**	*cis*	0.12	>30	**30**	*trans*	0.0096	>30
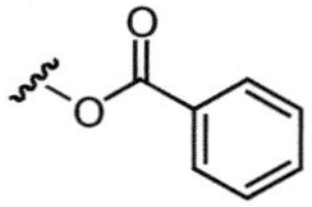	**31**	*cis*	0.049	>30	**32**	*trans*	0.11	>30
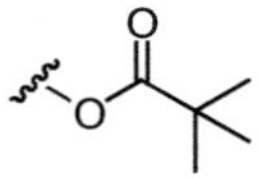	**33**	*cis*	0.54	>100	**34**	*trans*	0.019	>30
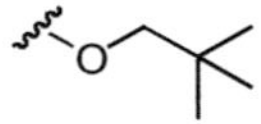	**35**	*cis*	17.0	>30	**36**	*trans*	0.18	>100
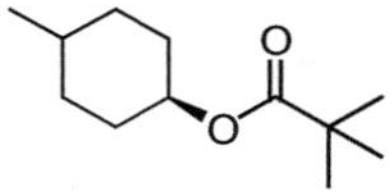	**37**	*cis*	>30	>30	**38**	*trans*	>30	>30

4.3 Isocyanobenzenes

As described above, isocyanocyclohexanes were shown to be promising non-toxic antifouling agents. However, the yields of these compounds were poor, because the generation of stereoisomers occurs during synthesis. To overcome this problem, the synthesis of isocyanobenzenes was performed. The isocyanobenzene with the isoprenyl side chain **39**, a phenyl version of 3-isocyanothonellin (**1**), was synthesized and found to be highly active against barnacle settlement – EC_{50} of 0.078 μg/mL (Table 5; Fig. 5) – without significant toxicity. This result suggests that isocyanobenzenes are promising as a new type of antifoulant. Therefore, structurally simpler isocyanobenzenes were prepared. 4-Alkyl isocyanobenzenes **40–42**, having a simple alkyl group at the C-4 position, showed moderate antifouling activity. Importantly, 4-benzyloxyisocyanobenzene (**43**) showed not only potent antifouling activity (EC_{50} = 0.054 μg/mL), but also high toxicity, as potent as that of $CuSO_4$. Moreover, these isocyanobenzenes could be prepared in one step from the corresponding comercially available aniline compounds. Thus, isocyanobenzenes are also thought to hold promise as antifouling agents.

Table 5. Structures of isocyanobenzenes **39-43** and their antifouling activity against cyprids of *B. amphitrite* after 120-h exposure (EC_{50} and LD_{50}). The antifouling activity of compounds is expressed as an EC_{50} value, which defines the concentration that reduces the larval settlement to 50% of the control. The toxicity of compounds is expressed as an LC_{50} value, which defines the concentration that results in 50% mortality.

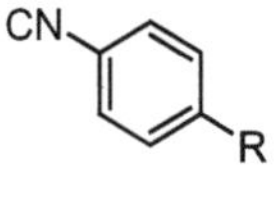

Compound No.	R	EC_{50} (μg/ml)	LD_{50} (μg/ml)
39	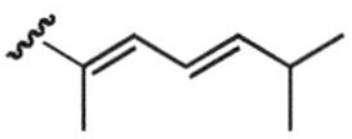	0.078	>100
40		0.64	14.2
41	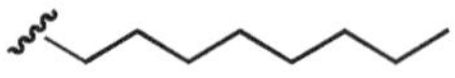	1.03	>3.0
42	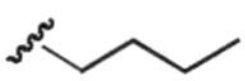	0.58	>100
43	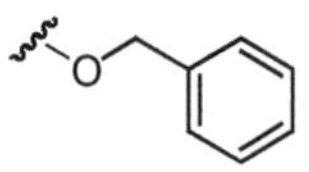	0.054	3.0

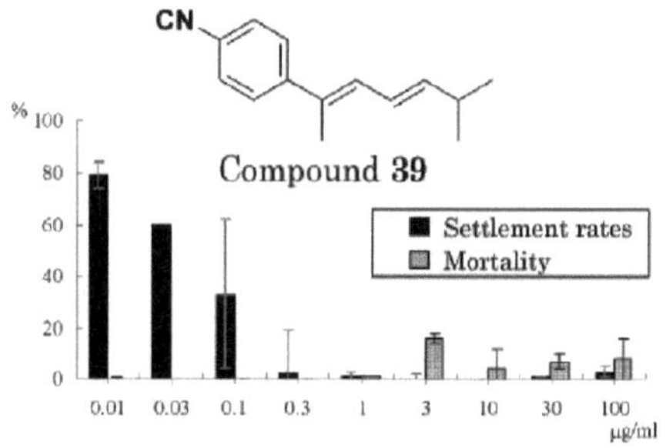

Fig. 5. Effects of synthesized isocyanobenzene **39** on settlement and mortality of cypris larvae after 120 h exposure. Rate of settlement of cyprids and mortality in different concentrations are shown as means ± SD of 3–5 replicates

4.4 Simple Linear Alkyl Isocyanides

In order to study further the structure-activity relationships, simple linear alkyl isocyanides were prepared (Nogata et al. 2004). Their structures, antifouling activity and toxicity are displayed in Table 6. Most of the synthesized linear alkyl isocyanides showed potent antifouling activity without significant toxicity. In particular, the alkene **44** and the phenyl sulfide **47** showed extremely strong activity, and their EC_{50} values were less than those of $CuSO_4$ and 3-isocyanotheonellin (**1**) (Fig. 6). Both antifouling activity and toxicity of the alcohol **45** were much higher than those of the phenyl esters **48–50**. The amine **46**, amide **51**, and imide **52** showed potent antifouling activity with EC_{50} values ranging from 0.10 to 0.21 μg/mL. However, the primary amine **46** and amide **51** were highly toxic to cyprids. This may be due to the presence of a hydrogen-bond donor. Di-isocyanide **53** also inhibited cyprid settlement at low concentrations without significant toxicity. It should be noted that the tertiary isocyanide **44**, which is mainly contained in natural isocyanoterpenes, showed more potent activity than the primary **54** and secondary isocyanide **55**. Perhaps the stability of the isocyano group affects the expression of antifouling activity.

4.5 A Large-Scale Synthesis of Isocyanide 44

Since simple, linear alkyl isocyanides showed potent antifouling activity without significant toxicity they are promising as environmentally benign antifouling agents. However, the supply issue must be solved for field experiments and the development of non-toxic antifouling agents. Candidate compound **44** can be synthesized from commercially available

Table 6. Structures of linear isocyanides **44–55** and their antifouling activity against cyprids of *B. amphitrite* after 120-h exposure (EC_{50} and LD_{50}). The antifouling activity of compounds is expressed as an EC_{50} value, which defines the concentration that reduces the larval settlement to 50% of the control. The toxicity of compounds is expressed as an LC_{50} value, which defines the concentration that results in 50% mortality

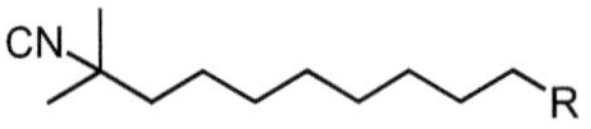

Compound No.	R	EC_{50} (μg/ml)	LD_{50} (μg/ml)
44	$CHCH_2$	0.046	>30
45	CH_2OH	0.31	>10
46	CH_2NH_2	0.16	21.3
47	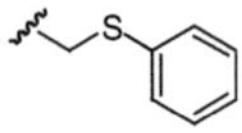	0.056	>30
48	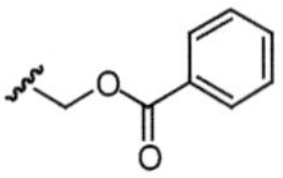	1.09	>300
49	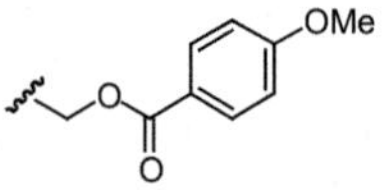	1.90	>300
50	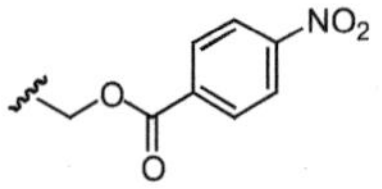	1.66	>300
51	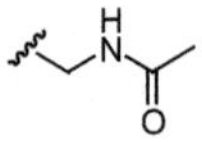	0.10	22.3
52	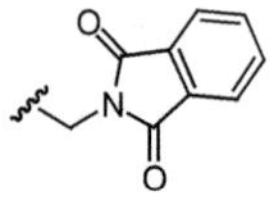	0.21	>100
53	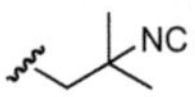	0.11	>100
54	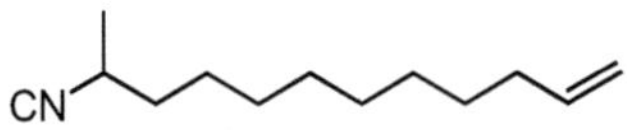	0.48	>100
55		0.14	70.0

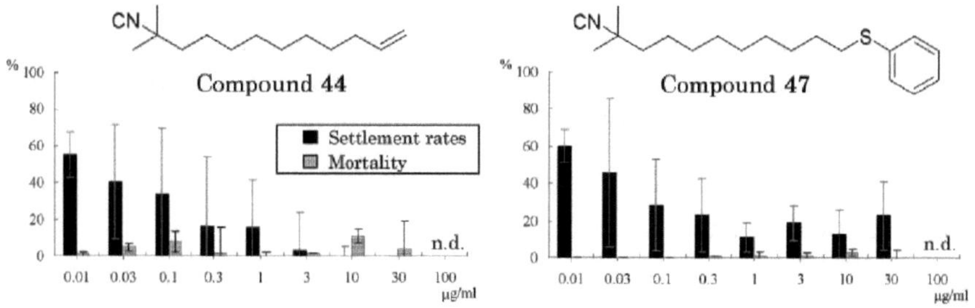

Fig. 6. Effects of synthesized linear alkyl isocyanides 44 and 47 on settlement and mortality of cypris larvae after 120 h exposure. Rate of settlement of cyprids and mortality in different concentrations are shown as means ± SD of 3–5 replicates

compound **56** in two steps as illustrated in Fig. 7. The construction of the isocyano functional group from alcohol, which is performed in the conversion of compound **56** to **57**, is a useful method to prepare isocyanides (Kitano et al. 1998, 2000).

5
Field Experiments with Isocyanide 44

In order to examine whether synthesized isocyanides prevent the settlement of fouling organisms in the field, the isocyanide 44 was incorporated into paint and evaluated for antifouling activity.

5.1
Test Panel Preparation

The test paint (100 g) was prepared by mixing compound 44 (15 g), acrylic copolymer including carboxylic acid (10.8 g), rosin (5.5 g), tricresyl phosphate (2.0 g), and titanium oxide (27.5 g) in a solution of 2-acetoxy-1-metoxy propane (39.2 g). After stirring the paint for 4 h, it was immediately applied onto PVC plates by brush.

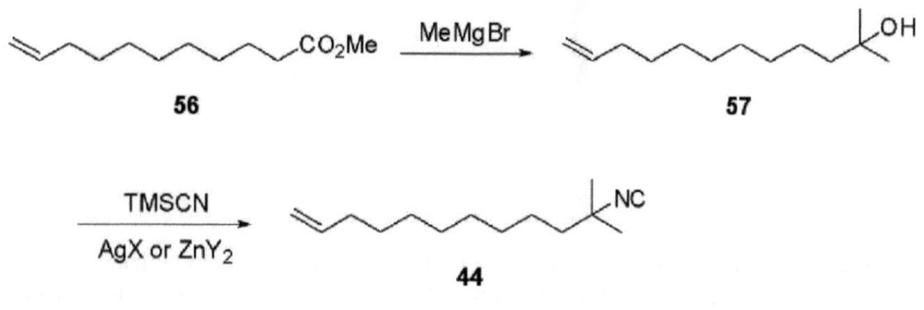

Fig. 7. Synthesis of compound 44

Gray PVC panels (25 cm × 25 cm × 3 mm) were sandblasted and then coated with the test paint and a copper-based paint (Figs. 8, 9). The paints were applied to only the front side of the panels, which were then left to dry in dark, in order to avoid photooxidation of the paint.

5.2 Field Experiments

5.2.1 *Field Experiment in Shizugawa Bay*

Field experiments were conducted at Shizugawa Bay, Miyagi, Japan (38° 38' N; 141° 27' E) between 30 August and 27 November 2003. The test panel (1 plate) was placed vertically at a depth of 0.5 m with ropes from a fishing boat. The test plate was evaluated monthly and photographed weekly during the test period.

Figure 8 shows the condition of the painted surfaces. The variability of fouling organisms on submerged test paints between sampling dates is shown in Table 7. The copper-based paint, which was lightly fouled with diatoms after 2 months, otherwise showed excellent antifouling performance during the test period. The trial paint showed antifouling activity against sessile organisms for over a month, although diatoms were found on the paint. Macrofouling organisms, such as ascidians and bryozoans, started to settle on the unpainted surface (backside) within 2 months, and thereafter they grew to cover all the surface by the end of

Fig. 8. Surface condition of test panels after an 86-day exposure in Shizugawa Bay. **A** Coated with copper-based paint; **B** coated with test paint; **C** non-painted control surface (*rear of test panel*)

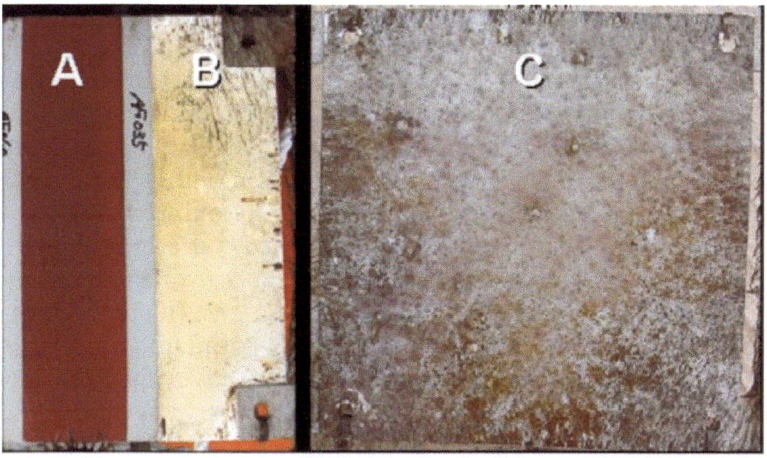

Fig. 9. Surface conditions of test panels after a 54-day exposure in Tokyo Bay. **A** Coated with copper-based paint; **B** coated with test paint; **C** non-painted control surface

Table 7. Abundance and timing of appearance of major fouling organisms on test paints in Shizugawa Bay

Organisms	September 30[1]		October 24		November 27		
	A[2]	B	A	B	A	B	C
Attached diatoms	-[3]	+[3]	+++	+	+++	+++	++
Algae	-	-	-	+	+	-	-
Hydrozoans	-	-	-	-	+	-	++
Bryozoans	-	-	-	-	-	-	++
Amphipods	-	-	-	-	-	-	+
Solitary ascidians	-	-	-	-	-	-	++
Compound ascidians	-	-	-	-	-	-	++

[1]Observation date
[2]Surface type: **A**, copper-based paint; **B**, the test paint; **C**, control (backside).
[3]Abundance of organisms: +++, very common (covered over 50%); ++, common (covered 10~50%); +, present; ±, rare (a few individual observed); –, absence

experiment (Fig. 8). On the other hand, the trial paint prevented settlement of ascidians and bryozoans during the experimental period. Compared with the unpainted surface, the trial paint retained antifouling activity for 3 months. Barnacles and oysters were observed alongside the pier wall, but were not found on the trial paint during the test period.

5.2.2
Field Experiments in Tokyo Bay

Field experiments were also carried out at Tokyo Bay (35° 37' N; 139° 46' E) between 16 October and 9 December 2003. The test panel (1 plate) was placed vertically from the pier at a depth of 1.0 m below the low-tide level.

The test plate was photographed on 12 November and 9 December (Fig. 9) and the nature and abundance of the fouling was noted (Table 8).

The copper-based paint completely prevented settlement of sessile organisms during the test period. Barnacles, *Balanus eburneus*, were apparent on the unpainted surface (back of plate) after one month of immersion, and ascidians had appeared by 2 months. The trial paint prevented settlement of ascidians completely, and reduced barnacle settlement in comparison with the unpainted surface. Although the painted surface was covered with diatoms and hydrozoans, only six barnacles (3.2 ind./100 cm^2) were found in 2 months. In comparison, 202 barnacles settled on the unpainted surface (32.3 ind./100 cm^2). It seems that trial paint retained promising anti-barnacle activity for a few months in the field.

Table 8. Abundance and timing of appearance of major fouling organisms on test paints in Tokyo Bay

Organisms	November 12[1]		December 9		
	A[2]	B	A	B	C
Attached diatoms	-[3]	+++[3]	-	+++	+++
Hydrozoans	-	-	+	-	++
Barnacles (No.)[4]	± (3)	-	± (6)	-	+ (202)
Solitary ascidians	-	-	-	-	+

[1]Observation date
[2]Surface type: A, copper-based paint; B, the test paint; C, control (backside).
[3]Abundance of organisms: +++, very common (covered over 40%); ++, common (covered 10~40%); +, present; ±, rare (a few individual observed); –, absence.
[4]Number of individuals observed

Table 9. The LD_{50}: EC_{50} ratios of synthesized isocyanides against cypris larvae of the barnacle *Balanus amphitrite.*

Compund No.	EC_{50} (μg/mL)	LD_{50} (μg/mL)	LD_{50}/EC_{50}
1	0.19	> 100	> 526.3
30	0.096	> 30	> 312.5
31	0.049	> 30	> 612.2
34	0.019	> 30	> 1578.9
39	0.078	> 100	> 1282.1
43	0.054	3.0	55.6
44	0.046	> 30	> 652.2
47	0.056	> 30	> 535.7
53	0.11	> 100	> 909.1
$CuSO_4$	0.30	2.95	9.8

6 Conclusion

In summary, isocyano compounds showed potent antifouling activity against cypris larvae, and some were not only more effective than cupric sulfate, but also significantly less toxic to cyprids. Marine isocyanides thus appear to be promising candidate non-toxic antifouling agents. However, the most potent natural products are often structurally too complex to be synthesized on a large scale (Clare 1996; Rittschof 2001). Hence, the synthesis was attempted of a variety of isocyano and related compounds based on 3-isocyanotheonellin. A number of the synthesized isocyanides showed potent antifouling activity without significant toxicity and hold promise as environmentally benign antifouling agents. In particular, linear alkyl isocyanides were strongly antifouling and could be synthesized easily. The ratio LD_{50}/EC_{50} indicates the relative effectiveness of the compound (Rittschof et al. 1992; Clare at al. 1999; Lau and Qian 2000). The LD_{50}/EC_{50} values of almost all isocyano compounds were approximately 10^2 or more; and the ratio for some compounds exceeded 10^4 supporting a non-toxic mechanism of action (Table 9). The field test results showed that a test paint incorporating an isocyanide prevented settlement of barnacles, ascidians and bryozoans.

The development of antifouling paints containing isocyanides will require a number of issues to be addressed including coating compatibility, controlled release and field testing. With respect to environmental acceptability, it will be necessary to determine whether the compounds biodegrade to biologically inactive products and ultimately that there are no effects on non-target organisms. The promising results obtained with initial trials suggest that isocyanides are very useful model compounds for the development of environmentally benign antifouling paints.

References

Abarzua S, Jakubowski S (1995) Biotechnological investigation for the prevention of biofouling. I. Biological and biochemical principles for the prevention of biofouling. Mar Ecol Prog Ser 123:301–312

Bakus GJ, Targett NM, Schulte B (1986) Chemical ecology of marine organisms: an overview. J Chem Ecol 12:951–987

Blunt JW, Copp BR, Munro MHG, Northcote PT, Prinsep MR (2004) Marine natural products. Nat Prod Rep 21:1–49

Clare AS (1996) Marine natural product antifoulants: status and potential. Biofouling 9:211–229

Clare AS, Rittschof D, Gerhart DJ, Hooper IR, Bonaventura J (1999) Antisettlement and narcotic action of analogues of diterpene marine natural product antifoulants from octocorals. Mar Biotechnol 1:427–436

Engel S, Pawlik JR (2000) Allelopathic activities of sponge extracts. Mar Ecol Prog Ser 207:273–281

Faulkner DJ (2002) Marine natural products. Nat Prod Rep 19:1–48 and previous reports in this series
Fusetani N (2004) Biofouling and antifouling. Nat Prod Rep 21:94–104
Fusetani N, Hirota H, Okino T, Tomono Y, Yoshimura E (1996) Antifouling activity of isocyanoterpenoids and related compounds isolated from a marine sponge and nudibranchs. J Nat Toxins 5:249–259
Garson MJ, Simpson JS (2004) Marine isocyanides and related natural products – structure, biosynthesis and ecology. Nat Prod Rep 21:164–179
Harper MK, Bugni TS, Copp BR, James RD, Lindsay BS, Richardson AD, Schnabel PC, Tasdemir D, vanWagoner RM, Verbitski SM, Ireland CM (2001) Introduction to the chemical ecology of marine natural products. In: McClintock JB, Baker BJ (eds) Marine chemical ecology. CRC Press, Boca Raton, pp 3–72
Hirota H, Tomono Y, Fusetani N (1996) Terpenoids with antifouling activity against barnacle larvae from the marine sponge *Acanthella cavernosa*. Tetrahedron 52:2359–2368
Hirota H, Okino T, Yoshimura E, Fusetani N (1998) Five new antifouling sesquiterpenes from two marine sponges of the genus *Axinyssa* and the nudibranch *Phyllidia pustulosa*. Tetrahedron 54:13971–13980
Kitano Y, Chiba K, Tada M (1998) A direct conversion of alcohols to isocyanides. Tetrahedron Lett 39:1911–1912
Kitano Y, Chiba K, Tada M (2000) Highly efficient conversion of alcohols to isocyanides. Synthesis 437–443
Kitano Y, Ito T, Suzuki T, Nogata Y, Shinshima K, Yoshimura E, Chiba K, Tada M, Sakaguchi I (2002) Synthesis and antifouling activity of 3-isocyanotheonellin and its analogues. J Chem Soc Perkin Trans 1:2251–2255
Kitano Y, Yokoyama A, Nogata Y, Shinshima K, Yoshimura E, Chiba K, Tada M, Sakaguchi I (2003) Synthesis and anti-barnacle activities of novel 3-isocyanotheonellin analogues. Biofouling 19 [Suppl]:187–192
Kitano Y, Nogata Y, Shinshima K, Yoshimura E, Chiba K, Tada M, Sakaguchi I (2004) Synthesis and anti-barnacle activities of novel isocyanocyclohexane compounds containing an ester or an ether functional group. Biofouling 20:93–100
Lau SCK, Qian PY (2000) Inhibitory effect of phenolic compounds and marine bacteria on larval settlement of the barnacle *Balanus amphitrite* Darwin. Biofouling 16:47–58
Nogata Y, Yoshimura E, Shinshima K, Kitano Y, Sakaguchi I (2003) Antifouling substances against larvae of the barnacle *Balanus amphitrite* from the marine sponge, *Acanthella cavernosa*. Biofouling 19 [Suppl]:187–192
Nogata Y, Yoshimura E, Shinshima K, Kitano Y, Sakaguchi I (2004) Antifouling activity of synthetic simple isocyanides against the barnacle *Balanus amphitrite* larvae. Biofouling 20:87–91
Okino T, Yoshimura E, Hirota H, Fusetani N (1995) Antifouling kalihinenes from the marine sponge *Acanthella cavernosa*. Tetrahedron Lett 36:8637–8640
Okino T, Yoshimura E, Hirota H, Fusetani N (1996a) New antifouling sesquiterpenes from four nudibranchs of the family Phyllidiidae. Tetrahedron 52:9447–9454
Okino T, Yoshimura E, Hirota H, Fusetani N (1996b) New antifouling kalihipyrans from the marine sponge *Acanthella cavernosa*. J Nat Prod 59:1081–1083
Omae I (2003) General aspects of tin-free antifouling paints. Chem Rev 103:3431–3448
Pawlik JR (1993) Marine invertebrate chemical defenses. Chem Rev 93:1911–1922
Rittschof D (2001) Natural product antifoulants and coatings development. In: McClintock JB, Baker BJ (eds) Marine chemical ecology. CRC Press, Boca Raton, pp 543–566
Rittschof D, Clare AS, Gerhart DJ, Avelin M Sr, Bonaventura J (1992) Barnacle *in vitro* assays for biologically active substances: toxicity and settlement inhibition assays using mass cultured *Balanus amphitrite amphitrite* Darwin. Biofouling 6:115–122
Thompson JE, Walker RP, Faulkner DJ (1985) Screening and bioassays for biologically-active substances from forty marine sponge species from San Diego, California, USA. Mar Biol 88:11–21

3-Alkylpyridinium Compounds as Potential Non-Toxic Antifouling Agents

K. Sepčić, T. Turk

Abstract. To date, around thirty bioactive 3-alkylpyridinium compounds, either in monomeric or oligomeric forms, have been identified in marine sponges belonging to the order Haplosclerida. In this work, we have reviewed their biological activities, which include mainly cytotoxicity, ichthyotoxicity, inhibition of bacterial growth, and enzyme inhibition. Most of these activities increase with the increasing degree of oligomerization of the corresponding 3-alkylpyridinium compound. It was shown recently that 3-alkylpyridines also exhibit promising antifouling activities. Linear 3-octylpyridinium polymers (Poly-APS), isolated from the Mediterranean sponge *Reniera sarai*, showed a non-toxic reversible mechanism of settlement inhibition of *Balanus amphitrite* cypris larvae with an EC_{50} of 0.27 µg/mL. At the same time, their toxicity towards the organisms used in the toxicity bioassays (*B. amphitrite* nauplii, microalga *Tetraselmis suecica* and larvae of *Mytilus galloprovincialis*) was almost negligible in comparison to commercially available and currently used booster biocides based on copper and zinc complexes with pyrithione. Poly-APS and some other natural 3-alkylpyridines were also found to be very effective in preventing microbial biofilm formation. Preliminary tests have confirmed that some monomeric and oligomeric synthetic analogues of poly-APS also exert antifouling activity, which makes these compounds promising candidates as new environmentally-friendly ingredients in the new generation of antifouling coatings.

1 Introduction

Sponges (Porifera) employ different defense strategies to survive in the highly competitive marine environment, and chemical defense is the one most frequently used. These sessile organisms produce a tremendous diversity of highly active compounds which exert a broad spectrum of biological activities – antibacterial, antifouling, cytotoxic, anti-inflammatory, antiviral and hemagglutinating activities, and enzyme inhibition. It is thus not surprising that recent trends in drug discovery have emphasized investigations of natural products from sponges. However,

K. Sepčić, T. Turk
Department of Biology, Biotechnical Faculty, University of Ljubljana, Večna pot 111, 1111 Ljubljana, Slovenia

Progress in Molecular and Subcellular Biology
Subseries Marine Molecular Biotechnology
N. Fusetani, A.S. Clare (Eds.): Antifouling Compounds

due to the high toxicity of many of the components, none of the compounds isolated in the last 20 years have reached the pharmaceutical marketplace (Faulkner 2000). In recent years, sponges have received increasing scientific attention on account of their specific skeleton structures, showing a potential use in the field of nanotechnology (Cha et al. 1999; Aizenberg et al. 2004). Finally, the increased search for environmentally-friendly antifouling compounds has rekindled interest in natural products having this particular activity (Rittschof 2000). As a consequence, several sponges have been tested and found to possess compounds exerting antifouling effects (Fusetani 2004).

Here, we review a group of sponge secondary metabolites bearing a 3-alkylpyridinium (3-AP) moiety, with special emphasis on polymeric alkylpyridinium salts (poly-APS) isolated from the Mediterranean sponge *Reniera (Haliclona) sarai* (Pulitzer-Finali). The range of biological activities of poly-APS has recently suggested these compounds as promising candidates for medicinal use as gene carriers (Tucker et al. 2003) or as non-toxic antifouling agents (Faimali et al. 2003a), thus justifying the efforts that are currently being dedicated to the chemical synthesis of their analogues (Mancini et al. 2004).

2 Origin, General Characteristics and Biological Activities of 3-Alkylpyridinium Compounds

Chemically related secondary metabolites often appear in taxonomically related sponges, e.g., bromotyrosines in the order Verongida, and guanidine and 2-aminoimidazole alkaloids in the genus *Agelas* (Almeida and Berlinck 1997), thus playing the role of chemotaxomic markers. During the last 30 years, a number of 3-AP compounds have been isolated from marine sponges belonging to the order Haplosclerida, suggesting that these compounds may serve as chemical markers for the systematic determination of haplosclerid sponges (Sepčić 2000; Tsukamoto et al. 2000). However, not all haplosclerid sponges contain 3-AP compounds. For example, only one species of *Haliclona* out of five collected in the Adriatic Sea contains 3-AP (T. Turk, unpubl. data). The occurrence of 3-AP compounds in different sponge genera is shown in Table 1. 3-APs were isolated from haplosclerid sponges either as (i) monomers differing in the length, saturation, branching and termination of the alkyl chains, (ii) cyclic or linear oligomers, or (iii) a mixture of high-molecular weight polymers. Despite their relatively simple chemical structure, all these compounds exert a broad spectrum of biological activities, which are listed and discussed below.

Table 1. Occurrence of 3-alkylpyridinium compounds in haplosclerid sponge genera

Class	Subclass	Order	Family	Genus	Compound	Reference
Non-identified sponge					ikimines	Carroll and Scheuer (1990)
Demospongiae	Tetractinomorpha	Lithistida	Theonellidae	*Theonella*	theonelladins	Kobayashi et al. (1989)
	Ceractinomorpha	Haplosclerida	Chalinidae	*Haliclona*	halitoxin	Schmitz et al. (1978)
					haliclamines	Fusetani et al. (1989)
					cyclostellettamines	Fusetani et al. (1994)*[a]
					viscosamine	Volk and Köck (2003)
					viscosaline	Volk and Köck (2004)
				Reniera	poly-APS	Sepčić et al. (1997a)
			Niphatidae	*Amphimedon*	amphitoxin	Albrizio et al. (1995); Kelman et al. (2001)
					halitoxin	Berlinck et al. (1996); Kelman et al. (2001)
					pyrinodemins	Tsuda et al. (1999); Hirano et al. (2000)
					unnamed 3-AP monomers	Hirano et al. (2000)
					hachijodines	Tsukamoto et al. (2000)
				Cribrochalina	cribrochalinamine oxides	Matsunaga et al. (1993)
				Niphates	niphatynes	Quiñoà and Crews (1987)
					niphatesines	Kobayashi et al. (1990, 1992)
					niphatoxins	Talpir et al. (1992)
					ikimine A	Kobayashi et al. (1992)
			Callyspongidae	*Callyspongia*	EGF-active factors	Davies-Coleman et al. (1993)
					untenines	Wang et al. (1996)
					halitoxin	Scott et al. (2000)
			Phloeodictydae	*Calyx*	xestamines	Stierle and Faulkner (1991)
			Petrosiidae	*Xestospongia*	xestamines	Sakemi et al. (1990)
					cyclostellettamines	Oku et al. (2004)
					hachijodines	Tsukamoto et al. (2000)

*[a] The sponge was initially identified as *Stelletta maxima*; however re-examination of the specimen has revealed that cyclostellettamines were contained in an epiphytic *Haliclona* sp. (Oku et al. 2004)

3 Monomeric 3-Alkylpyridinium Compounds

More than 20 monomeric 3-APs have been isolated from marine sponges of the order Haplosclerida and the majority of them bear a nitrogenous functionality at the end of the alkyl chain. The first to be reported were niphatynes, bearing alkyne and the *N*-methoxyamine functionalities (Quiñoà and Crews 1987). They were followed by theonelladins terminating in the primary amino or methylamino group (Kobayashi et al. 1989) and ikimines terminating in the oximino or *N*-methoxyamino group (Carroll and Scheuer 1990; Kobayashi et al. 1992). Aliphatic 3-AP with the *N*-methoxy-*N*-methylamino terminus, xestamines, were reported from two haplosclerid sponges (Sakemi et al. 1990; Stierle and Faulkner 1991), while a sponge of the genus *Niphates* was found to possess a variety of niphatesines bearing the amino (Kobayashi et al. 1990), *N*-metoxyamine, or oxime methyl ether functionality at the end of the alkyl chain (Kobayashi et al. 1992). Cribrochalinamine oxides possess an azomethine *N*-oxide function in the side chain (Matsunaga et al. 1993), untenines were isolated as nitroalkyl 3-alkylpyridines (Wang et al. 1996), hachijodines terminate in the *N*-metoxyamino or *N*-hydroxy-*N*-methylamino groups (Tsukamoto et al. 2000), and unnamed 3-alkylpyridines from *Amphimedon* sp. in the oxime group (Hirano et al. 2000).

Almost all reported 3-AP monomers exert moderate cytotoxicity at low microgram per millilitre concentrations against certain transformed cell lines. Surprisingly, xestamines A-C from *Xestospongia widenmayeri* (Sakemi et al. 1990) were inactive against P-388 cells *in vitro*, despite their structural resemblance with niphatyne A which has an IC_{50} of 0.5 μg/mL against the same cell line (Quiñoà and Crews 1987). Further isolation of xestamines D-H from *Calyx podatypa* yielded two fractions, xestamines and their corresponding *N*-methylpyridinium salts. It is interesting that these quaternary pyridinium salts exerted a moderate antibacterial activity against G^+ bacteria that was about 40 times higher than the activity of the corresponding tertiary amines. In contrast, the latter showed about 30 times higher activity than the quaternary pyridinium salts in the brine shrimp cytotoxicity assay (Stierle and Faulkner 1991). Other reported biological effects induced by 3-AP monomers include antibacterial (Kobayashi et al. 1992; Hirano et al. 2000) and antifungal activities (Kobayashi et al. 1992; Matsunaga et al. 1993; Hirano et al. 2000), antimicrofouling activity against the marine bacterium *Phodospirillum salexigens* with IC_{100} values of 3.0–6.1 μg/mL (Wang et al. 1996), and Ca^{2+}-releasing activity from the sarcoplasmic reticulum (Kobayashi et al. 1989).

4 Dimeric and Trimeric 3-Alkylpyridinium Compounds

The majority of 3-AP dimers and trimers were isolated from haplosclerid sponges in cyclic forms. Two different structural types of cyclic dipyridines have been reported so far: haliclamines – tetrahydropyridines linked through C_9 and C_{12} alkyl chains (Fusetani et al. 1989) – and cyclostellettamines, which consist of two pyridines linked by C_{12} to C_{14} alkyl chains (Fusetani et al. 1994; Oku et al. 2004). Haliclamines A and B inhibited the division of fertilized sea urchin eggs, as well as the growth of transformed cell lines (Fusetani et al. 1989), while cyclostellettamines A-F were reported to inhibit muscarinic acetylcholine receptors (Fusetani et al. 1994), as well as histone deacetylase enzymes that are being considered as therapeutic targets for treating cancer, and the growth of transformed cell lines (Oku et al. 2004). Examples of linear 3-AP dimers are pyrinodemins, bis-pyridine alkaloids with a unique isoxazolidine moiety (Tsuda et al. 1999; Hirano et al. 2000). It is interesting that these linear 3-AP dimers showed 10 to 100 fold higher cytotoxicities against transformed cell lines than the 3-AP monomers. This activity is comparable to that of the only reported linear 3-AP trimers, niphatoxins (Talpir et al. 1992). A cyclic trimeric 3-AP compound, viscosamine, was isolated from the Arctic haplosclerid sponge *Haliclona viscosa* (Volk and Köck 2003). Viscosaline, a 1,3-dialkylpyridinium compound with a β-alanine moiety covalently bound to one alkyl chain, was recently purified from the same sponge (Volk and Köck 2004).

5 Polymeric 3-Alkylpyridinium Compounds

The isolation of 3-AP polymers from haplosclerid sponges is rather problematic, since they usually exist as a mixture of compounds with the same basic structure but with different molecular weights. Furthermore, the polarity of 3-APs increases with the degree of oligomerization, leading to the formation of non-covalently bound supramolecular aggregates and rendering the separation of different molecular weight oligomers and polymers even more difficult. The polymeric 3-APs reported to date are composed of monomeric subunits with either different (halitoxin, amphitoxin), or the same basic structure (EGF-active factors, poly-APS), as shown in Fig. 1.

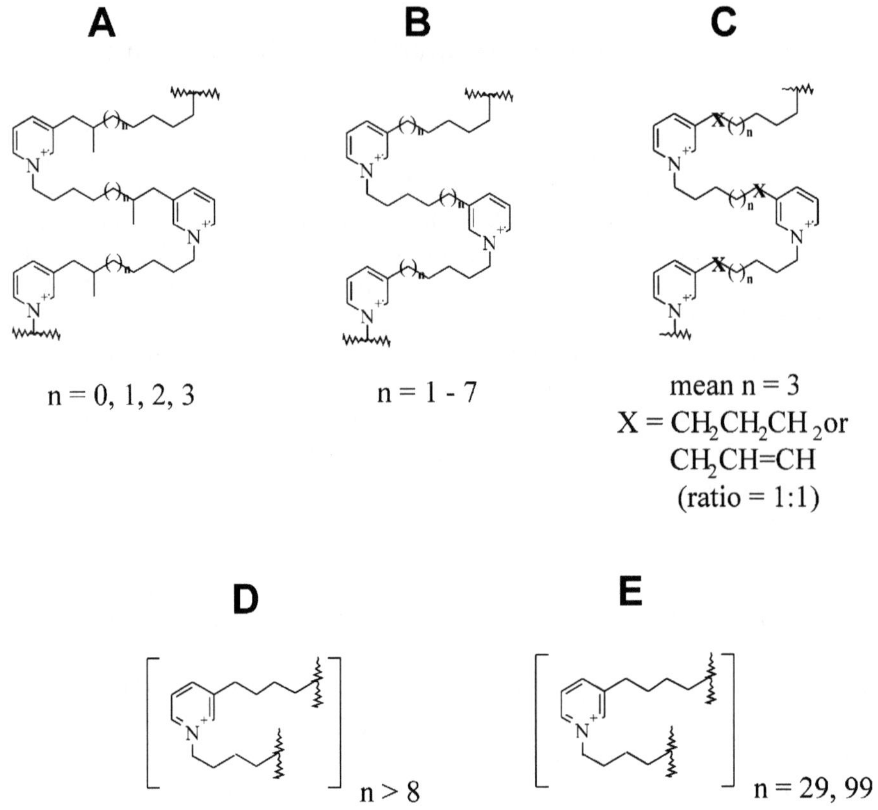

Fig. 1. Chemical structures of 3-alkylpyridinium polymers isolated from haplosclerid marine sponges. **A** Halitoxins (Schmitz et al. 1978); **B** halitoxins (Scott et al. 2000); **C** amphitoxins (Albrizio et al. 1995); **D** EGF-active factors (Davies-Coleman et al. 1993); **E** poly-APS (Sepčić et al. 1997a).

5.1 Halitoxin and Amphitoxin

Halitoxin, a mixture of high molecular weight pyridinium salts isolated from several sponges of the genus *Haliclona* (Schmitz et al. 1978), was the first 3-AP polymer structurally characterized. The crude toxin was roughly separated by membrane ultrafiltration into molecular weight range fractions of 500–1,000 (the major fraction), 1,000–25,000 and >25,000 Da, all having the same basic structure – 3-AP units connected with saturated methyl-branched alkyl chains composed of 8 to 11 carbon atoms (Fig. 1A). In 1996, Berlinck and co-workers also isolated halitoxin from the Brazilian sponge *Amphimedon (Haliclona) viridis*. The isolated compounds had the same basic structure as those isolated by Schmitz et al. (1978), but with a different distribution of molecular weights: 500, 2,000 (the major peak, corresponding to more than 90% of the complex) and 5,000 Da. Further structural characterization of unbranched halitoxins isolated from *Callyspongia ridleyi*, using ESI and MALDI-TOF mass spectrometry, suggested that each polymer chain was made up of a random sequence of monomers differing in the length of the unbranched saturated alkyl chains, and that the number of monomers in the polymer chain ranged from 19 to 27 (Scott et al. 2000) (Fig. 1B). Another high molecular weight pyridinium salt, amphitoxin, was isolated from the haplosclerid sponge *Amphimedon compressa* (Albrizio et al. 1995). Just like halitoxin, it is a polymer containing monomeric subunits of different sizes, but it differs in possessing unsaturated alkyl chains. Amphitoxin is based on randomly sequenced 3-alkyl and 3-alkenyl pyridinium units in an overall ratio of 1:1 (Fig. 1C). Purification of the amphitoxin mixture by membrane ultrafiltration gave fractions in two molecular weight ranges: 1,000–3,000 (38%) and 3,000–10,000 Da (62%). Recently, a mixture of two closely related polymeric 3-AP homologues, identified as halitoxin and amphitoxin, was isolated from *Amphimedon viridis*, but could not be separated (Kelman et al. 2001).

Halitoxins and amphitoxins, isolated so far, showed a broad spectrum of biological activities. The groups of both Schmitz and Berlinck reported on the haemolytic activity of halitoxins and their lethality to mice. These two activities increased with increasing molecular weight (Berlinck et al. 1996). Halitoxin was also cytotoxic to transformed cells, toxic to fish and mice, antibacterial towards *Bacillus subtilis* and *Streptococcus pyogenes* (Schmitz et al. 1978), antimitotic against fertilized sea urchin eggs, and neurotoxic in a crustacean nerve assay (Berlinck et al. 1996). Electrophysiological actions of halitoxins were studied in detail by Scott et al. (2000), using primary cultures of rat dorsal root ganglion neurons, as well as artificial lipid bilayers. In cell cultures, halitoxin caused irreversible membrane potential depolarization, decreased input resistance, and inhibited evoked action potentials. The toxin also evoked a calcium influx and its release from intracellular stores. Halitoxin also induced a channel-like activity in artificial lipid bilayers composed of phosphatidylcholine and cholesterol. At one-sixth of its natural concentration, amphitoxin showed a strong anti-feedant activity to the Caribbean fish *Thalassoma bifasciatum* (Albrizio et al. 1995). Kelman and co-workers (2001) have assessed the antimicrobial activity of the complex halitoxin/amphitoxin mixture that was isolated from *A. viridis*. The mixture was tested against ecologically relevant bacteria, which resulted in no activity against sponge-associated bacteria (which also had no match in bacterial databases of the biochemical kits that were used for the identification), while strong inhibition was found against most strains isolated from seawater. This selective inhibition could have an important ecological role in preventing microfouling and in defense against potentially pathogenic marine bacteria.

5.2 EGF-Active Factors

In addition to halitoxins and amphitoxins, which differ in the structure of their monomeric subunits, two 3-AP polymers composed of equal monomers were isolated from haplosclerid sponges. *Callyspongia fibrosa* was found to be a rich source of 3-AP polymers linked head-to-tail through straight C_8 alkyl chains and containing at least eight monomeric subunits (Davies-Coleman et al. 1993; Fig. 1D). These compounds were named EGF-active factors, as they were able to inhibit the receptor of the epidermal growth factor (EGF), which is overexpressed in several tumor cells. The authors have also tried to synthesize analogues of EGF-active factors using head-to-tail oligomerization. The procedure yielded a mixture of dimer, trimer, tetramer, and higher order oligomers. Because of their increasing polarity, further separation of higher order oligomers was prevented. Comparison of biological activities of different synthetic EGF-derivatives with those of the natural compound had again shown that the higher degree of polymerization was responsible for the increased biological activity. Later, 3-AP oligomers, mimicking the EGF-active factors and bearing an oxygen atom in place of a methylene group in the alkyl chain, were synthesized and tested for cytotoxic activity against KB cells (Gil et al. 1995). Again, oligomers having a higher degree of polymerization proved to be more active and soluble exclusively in water.

5.3 Polymeric Alkylpyridinium Salts (Poly-APS)

In 1997, polymeric 3-AP salts with the same basic structure as EGF active-factors, but with a higher degree of polymerization, were isolated from the Mediterranean sponge *Reniera (Haliclona) sarai* and named poly-APS (Sepčić et al. 1997a). Poly-APS show promising biological activities and are one of the most investigated 3-APs. Therefore, they are presented here in greater detail.

5.3.1 Isolation and Structural Characterization of Poly-APS

In contrast to other 3-AP derivatives, that have always been isolated from different organic sponge extracts, poly-APS were purified from a crude aqueous extract of *R. sarai*. MALDI-TOF spectrometry has shown that

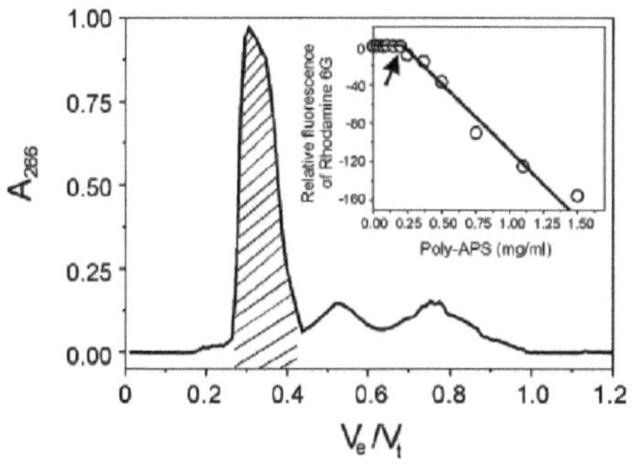

Fig. 2. Typical elution profile of the final purification step of poly-APS. Extracted poly-APS were eluted from a Sephadex G-50 column at the void volume (*dashed*). V_e = elution volume, V_t = total volume. *Inset*: Determination of the aggregation point of poly-APS using a fluorescent probe Rhodamine 6G as described by de Vendittis et al. (1981).

poly-APS are a mixture of two main polymers with molecular weights of 5520 and 18900 Da, corresponding to polymers composed of 29 and 99 3-octylpyridinium units, respectively (Fig.1E). Purified poly-APS are soluble only in water. Irrespective of their basic size, at concentrations of more than 230 µg/mL they form large supramolecular aggregates with an average hydrodynamic radius of 23±2 nm. Due to the high concentrations of poly-APS in the starting material, that exceeded the aggregation point of these compounds (as revealed by the use of fluorescent probe Rhodamine 6G), they eluted from the Sephadex G-50 column as a single peak at the void volume (Fig. 2). The aqueous extract of *R. sarai* also contains oligo-APS, 3-octylpyridinium oligomers with molecular weights considerably lower than 3,000 Da and without the significant biological activities associated with poly-APS (Sepčić et al. 1997a).

5.3.2
The Biological Activities of Poly-APS

Poly-APS are strong inhibitors of acetylcholinesterase (AChE) (Sepčić et al. 1997a), and show moderate haemolytic and cytotoxic activity (Sepčić et al. 1997b). Their IC_{50} towards transformed cell lines is approximately 0.3 µg/mL, which is about 10 times lower than that of halitoxin (Schmitz et al. 1978). The haemolytic activity of poly-APS is probably due to their detergent-like properties. Experiments using osmotic protectants of various sizes suggested that poly-APS induce a colloid-osmotic type of lysis by producing discrete lesions, 5.8 nm in diameter, in cell membranes (Malovrh et al. 1999). The ability of poly-APS to induce transient pores in cell membranes (McClelland et al. 2003) was used to assess the transfection of human embryonic kidney cells (HEK 293) with plasmid cDNA (Tucker et al. 2003). As already reported for halitoxin (Scott et al. 2000), poly-APS induced a collapse in membrane potential, reduced input resistance and increased Ca^{2+} permeability in HEK 293 cells. In contrast to halitoxin, at least partial recovery was observed following poly-APS application. As a consequence, poly-APS have enabled stable transfection of living cells with double-stranded DNA, although they are 2.5 fold less efficient than lipofectamine (Tucker et al. 2003).

The anti-AChE activity of poly-APS has also been studied in detail (Sepčić et al. 1998). Pyridinium derivatives are well-known inhibitors of acetylcholinesterases; they act as competitive inhibitors, binding at the catalytic anionic site at the bottom of the enzyme gorge or, as non-competitive inhibitors, binding at the peripheral anionic site at the rim of the gorge, thus slowing or preventing the entrance of acetylcholine into the gorge. However, the kinetic of AChE inhibition by poly-APS is complex and comprises several successive phases ending in irreversible inhibition of the enzyme. The irreversible phase is probably accounted for by aggregation and precipitation of enzyme-inhibitor complexes (Sepčić et al. 1999). Comparison of poly-APS anticholinesterase activity with that of smaller pyridinium inhibitors shows that this activity, like the other biological activities described above, increases with increasing polymerization. For example, simple monomeric alkylpyridinium derivatives, like methyl- or ethylpyridines, act as pure competitive inhibitors (Whiteley and Ngwenya 1995), with an approximately 1,000 fold higher K_i than poly-APS.

Despite these potent effects of poly-APS observed *in vitro*, the inhibition of AChE was not found to be the principal factor for lethality of rats *in vivo*. Especially at higher poly-APS doses, AChE inhibition is masked by, or is secondary to other mechanisms leading to death, such as thrombocyte aggregation (Bunc et al. 2002a), plug formation in the tissues (Bunc et al. 2000), and non-specific binding to plasma proteins (Bunc et al. 2002b).

5.3.3
Antifouling Activity of Poly-APS

We became interested in the antifouling activity of poly-APS as a consequence of our observation that the Mediterranean marine sponge *Reniera sarai* is almost never fouled by other marine organisms (Fig. 3). Since the greasy surface of the sponge is evident, this characteristic can be easily associated with poly-APS, which might play an important role in the sponge's antifouling strategy. As described above, poly-APS exhibit detergent-like properties and, consequently, are capable of exercising a number of biological activities. The surfactant activity towards artificial and natural membranes and AChE inhibitory activity might be involved in antifouling molecular mechanisms that prevent settlement of bacteria and metazoa onto the sponge surface.

Fig. 3. *Reniera sarai* (Pulitzer-Finali). Photo: Tom Turk.

Inhibition by poly-APS of the settlement of *Balanus amphitrite* cypris larvae was tested. Stage II nauplii of *B. amphitrite*, together with some other ecologically relevant planktonic organisms (microalga *Tetraselmis suecica* and larvae of the edible mussel *Mytilus galloprovincialis*), were used in toxicity assays (Faimali et al. 2003a). In addition, poly-APS were tested for their potential anti-microfouling activity, for example their ability to prevent the formation of a biofilm on submerged surfaces under laboratory conditions (Garaventa et al. 2003).

The results were compared to those obtained with two booster biocides based on copper and zinc complexes with pyrithione, respectively. These booster biocides (Zinc Omadine® and Copper Omadine®) are commercially available and are currently used in certain antifouling coatings.

Antifouling activity of poly-APS, tested on *Balanus amphitrite* cypris larvae, showed an EC_{50} of 0.27 μg/mL, and was higher than the corresponding activity of the crude extract obtained from *R. sarai* (EC_{50} = 1.46 μg/mL). These values were higher than those obtained with metallic pyrithione compounds, that gave EC_{50}s for inhibition of cypris settlement of 0.01 (copper pyrithione) and 0.2 (zinc pyrithione) μg/mL. However, while booster biocides exhibit significant toxicity, even at these low concentrations, poly-APS at their EC_{50} do not show any significant toxicity, either against *B. amphitrite* nauplii in the swimming inhibition assay, or in the naupliar toxicity assay. Poly-APS were also considerably less toxic against *Mytilus galloprovincialis* larvae (veliger and trocophora) in a bivalve acute toxicity test. The effective values of poly-APS compared with booster biocides are summarized in Table 2.

The lack of toxicity associated with settlement inhibition by poly-APS was proved by the reversibility of the latter. Cypris larvae were exposed for 72 h to poly-APS at the concentration that caused 100% settlement inhibition. They were then rinsed and placed into fresh seawater, where their settlement was monitored. After 120 h the cyprids were able to settle at a rate not significantly different from that of untreated larvae. In contrast, larvae treated with booster biocides were completely unable to

Table 2. Comparison of EC_{50}, LC_{50} and IC_{50} values of poly-APS and booster biocides for different test organisms. EC_{50}, the concentration of the antifouling compound causing settlement inhibition of 50% of experimental organisms; IC_{50}, the concentration of the antifouling compound causing inhibitory effects (swimming inhibition or inhibition of algal duplication) on 50% of experimental organisms; LC_{50}, the concentration of the antifouling compound causing death of 50% of experimental organisms. Data represent median concentration ±95% confidence interval. NC = not calculated. All values are in μg/mL. (Published in Faimali et al. 2003a, with permission of Taylor & Francis)

Species	Bioassay	End-points	Poly-APS	Zinc Pyrithione	Copper Pyrithione
Balanus amphitrite	Cyprid Settlement Inhibition	24h EC_{50}	0.27 (0.47–0.15)	0.02 (NC)	<0.01
	Naupliar Mortality	24h LC_{50}	30.01 (41.49–21.71)	0.19 (0.30–0.13)	<0.01
	Naupliar Swimming Inhibition	24h IC_{50}	>10	0.23 (0.33–0.16)	0.03 (0.04–0.03)
Teraselmis suecica	Algal Growth Inhibition	72h IC_{50}	10.66 (NC)	0.29 (0.6–0.14)	0.024 (0.04–0.01)
Mytilus galloprovincialis	Trocophora Mortality	24h LC_{50}	2.42 (4.26–1.78)	0.004 (0.005–0.003)	0.012 (NC)
	Veliger Mortality	24h LC_{50}	2.73 (3.35–2.16)	0.017 (0.021–0.014)	0.036 (0.047–0.026)

settle, suggesting that the antifouling activity was due to their toxicity (Fig. 4).

In conclusion, although less effective than the booster biocides, poly-APS proved to be considerably less toxic. Thus, at least in an acute sense, the antifouling activity of poly-APS can be considered as non-toxic. Their inhibitory effect on *B. amphitrite* larvae is comparable to those obtained by some other compounds recently isolated from marine sponges and reviewed by Fusetani (2004) (Table 3).

The molecular mechanism by which poly-APS inhibit settlement of cyprids can be explained in several ways. These compounds are known to strongly inhibit AChE (Sepčić et al. 1998, 1999). Recently it was suggested that acetylcholine is involved in the settlement of *B. amphitrite* cyprids (Faimali et al. 2003b). In this study biochemical, histochemical and immunohistochemical data demonstrated the presence of a functional set of cholinergic molecules in *B. amphitrite* cyprids, predominantly in the thoracic appendages and the caudal rami extending from the cyprid's thoracic region. AChE activity was also detected in the setae of the antennules, which play an important role in the process of substratum recognition and subsequent settlement (Clare et al. 1994). It is therefore probable that AChE is involved in the larval settlement process. The raised level of acetylcholine, due to the inhibition of AChE by several agents, caused an increase in larval settlement, providing that at least some of the enzyme is still active. However, this effect was observed only at rather low concentrations of AChE inhibitors (≤ 1 μg/mL), while higher concentrations caused a clear inhibition of larval settlement. The

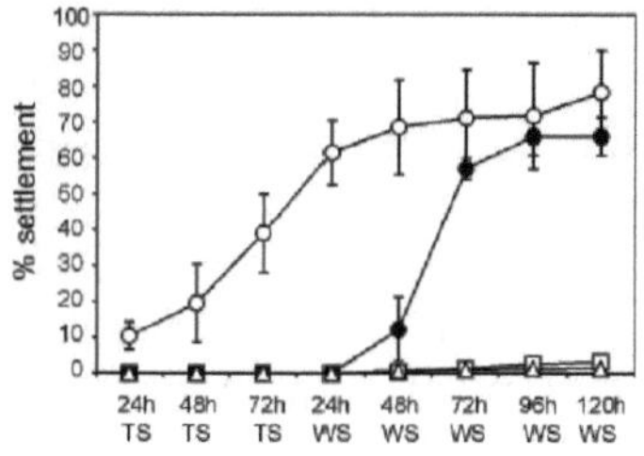

Fig. 4. Reversibility of settlement inhibition on *Balanus amphitrite* cyprids. *B. amphitrite* cyprids were exposed for 72 h to the test solution (TS) of poly-APS (●), zinc pyrithione (□) and copper pyrithione (Δ) at their 100% settlement inhibition concentrations. The cyprids were then rinsed with filtered natural seawater and their settlement ability was checked at 24, 48, 72, 96 and 120 h after rinsing (water solution; WS). In the control group (○), cyprids were kept in pure filtered natural seawater without biocides during the course of the experiment. Data = mean ±SD (n = 5). (Published in Faimali M, Sepčić K, Turk T, Geraci S (2003) Biofouling 19:47–56, with permission of Taylor & Francis).

Table 3. Antifouling activity and toxicity of some compounds obtained from different marine sponges

Sponge species	Trivial name	Chemical class	Antifouling activity IC_{50} (µg/mL) (*B. amphitrite*)	Toxicity LC_{50} (µg/mL) (*B. amphitrite*)
Acanthella cavernosa	Kalihinene X	Terpenoid	<0.1	>100
	Kalihipyran B		<0.1	>100
Axinyssa sp.	Axinyssamide A	Terpenoid	1.2	n.d.
	Axinyssamides B, C		<0.5	
Various sponges	Halistanol sulfate	Steroid	5.0	n.d.
Pseudoceratina purpurea	Ceratinamine	Bromotyrosine derivative	5.0	n.d.
	Ceratinamide A		0.1	
	Psammaplysin A		0.27	
Agelas mauritiana	Mauritiamine	Oroidine derivative	15.0	n.d.
Reniera sarai	Poly-APS	Alkylpyridinium polymer	0.27	30

n.d. = not determined

enhancement of settlement at low inhibitor concentrations might be explained by an overcompensating response to disruption of homeostasis, which results in an apparent low-dose stimulation effect (Calabrese and Baldwin 2001). However, at higher inhibitor concentrations, this compensation response is limited due to the complete inhibition of the enzyme. At this stage, no cyprids are able to settle on the surface. According to these results one could assume that poly-APS inhibit settlement of cyprids by inhibiting their cholinergic system.

An alternative reason for the observed settlement inhibition could derive from the surfactant-like characteristics of poly-APS, preventing attachment to the surface. Indeed, in addition to *B. amphitrite* larval settlement inhibition, poly-APS also inhibit microfouling by bacteria, fungi, and microalgae (Garaventa et al. 2003). In aqueous solutions, poly-APS behave similarly to detergents that are classified as quaternary ammonium compounds. According to their critical micelle concentration and charge, poly-APS resemble well-known cationic detergents like cetylpyridinium chloride and cetyltrimethylammonium bromide (Malovrh et al. 1999). These compounds are examples of synthetic surfactants used in microbial adhesion studies and are well-known antiseptic agents in toothpaste and mouth wash. Surprisingly, there is at least one study that reports positive adhesion effects on bacteria when cetylpyridinium chloride was bound to a negatively charged bacterial surface. This binding increased the hydrophobicity of bacteria and enhanced adhesion to the chosen surface (Goldberg et al. 1990). Poly-APS are lytic to different cell lines, but according to our results, they do not significantly inhibit growth of terrestrial bacteria (Sepčić et al. 1997b; Mancini et al. 2004). On the other hand, according to laboratory assays, these compounds are quite effective against marine bacteria. Poly-APS show inhibition of marine bacterial growth in the range of 0.1–1 µg/mL (Garaventa et al. 2003). Similar selective antibacterial activity was previously reported for the amphitoxin/halitoxin mixture (Kelman et al. 2001). In the latter study, the resistant bacterial strains were also resistant to polymyxin B, an antibiotic that acts by binding to the bacterial membrane. The antibiotic activity of polymeric 3-APs could follow the same mechanism, as their positive charge allows binding to the bacterial surface.

Furthermore, poly-APS also inhibit microfouling by fungi and algae, albeit at rather higher concentrations than booster biocides. They prevent settlement of diatoms and growth of microalga *Tetraselmis suecica* (Faimali et al. 2003a; Garaventa et al. 2003). The IC_{50} value for the growth of *T. suecica* was 10.66 µg/mL, a concentration that is about 450 times higher than that for the most effective booster biocide, copper pyrithione (Faimali et al. 2003a) (Table 2).

6 Ecological Significance of 3-Alkylpyridinium Compounds

Although considered as the most primitive multicellular organisms, sponges (Porifera) considerably surpass other sessile organisms in the production of chemically novel secondary metabolites. The lack of a locomotor system makes these animals an easy target for different predators, fouling by micro- and macroorganisms, and other sessile organisms participating in territorial competition. 3-alkylpyridinium compounds, produced by haplosclerid sponges in large quantities and showing a broad spectrum of biological activities, are most probably involved in sponge defense. Indeed, the surface of *Reniera sarai* is smooth and clean, veiled in a greasy secretion containing large amounts of poly-APS that probably act as a protective coating against fouling (T. Turk, pers. obs.). Similar surface characteristics have also been reported for other sponges containing polymeric 3-APs, e.g., in *Callyspongia* (Scott et al. 2000), and *Amphimedon* (Kelman et al. 2001). The selective activity of the halitoxin/amphitoxin complex against different marine bacteria, with inhibition of seawater strains and inactivity towards sponge-associated bacteria (Kelman et al. 2001), further suggests that polymeric 3-APs act as natural antifoulants. Strong anti-feedant activity of amphitoxin, observed with fish (Albrizio et al. 1995), suggests that these compounds might serve also as natural repellents and points to their protective role against predators. Finally, because of their ability to enable stable transfection of viable cells with double-stranded DNA, polymeric 3-APS might provide a natural mechanism for the exchange of genetic material between marine organisms (Tucker et al. 2003).

7 Perspectives

As for other natural compounds that might find application in commercial products, a major issue for the use of natural antifoulants is finding an environmentally responsible and friendly way to obtain adequate quantities of such compounds. There are several means of overcoming this problem that include aquaculture, cell culture, chemical synthesis and, in certain cases where proteins are involved, genetic manipulation and expression in bacteria. The most common approach to obtain large quantities of a natural product is total chemical synthesis or even synthesis of analogues that might have superior characteristics over a parent compound. This approach is usually successful, providing that the structure of the natural compound is not too complicated.

Several groups have reported successful synthesis of 3-AP compounds, such as monomeric niphatesines (Rao and Reddy 1993), or dimeric cyclostellettamines (Baldwin et al. 1998). However, the synthesis and separation of higher 3-AP oligomers can be rather difficult due to increasing polarity of the originating oligomers. Davies-Coleman and co-workers (1993) have used a traditional approach inducing head-to-tail oligomerization for the synthesis of the EGF-active 3-APs from the sponge *Callyspongia fibrosa*. They obtained a mixture of dimer, trimer, tetramer and higher order oligomers, but the latter were difficult to separate because of their increasing polarity. Similar problems were encountered by Gil et al. (1995). The same authors later realized the synthesis using the *N*-protection/C-activation sequence involving the Zincke reaction, allowing the number of alkylpyridinium subunits in the polymers to be controlled by an iterative process (Kaiser et al. 1998). Similarly, in order to obtain linear oligomeric analogues of poly-APS, we have adopted the *N*-protection/C-activation sequence as the method of choice (Mancini et al. 2004). Linear dimers and tetramers have been synthesized and tested for several biological activities, together with other 3-AP intermediates formed during the synthetic process.

Several authors have shown that polymeric 3-AP compounds have higher biological activity in most assays than oligomeric or monomeric

3-APs (Davies-Coleman et al. 1993; Gil et al. 1995; Berlinck et al. 1996; Mancini et al. 2004). This general rule was observed in the case of cytolytic and cytotoxic activities, and inhibition of AChE. The only notable exception was the antimicrobial activity against terrestrial bacteria, where oligomers showed superior activity over polymers (Mancini et al. 2004).

The reversible antifouling effects induced by poly-APS under laboratory conditions, together with their low toxicity, and their solubility and stability, make these compounds promising candidates for incorporation in antifouling paints and/or other protective coatings. It is very probable that other polymeric 3-APs (halitoxin, amphitoxin, EGF-active factors) could exert similar antifouling effects, and it would be interesting to see whether the antifouling activity and toxicity of these compounds depend of their molecular weight. Of course, a successful route for synthesizing high molecular 3-AP compounds would facilitate their *in situ* testing, as well as their potential future use as environmentally friendly antifouling agents. Some synthetic 3-carboxyamido pyridinium derivatives have already been assayed and found to be active against microfouling by marine bacteria (Răşanu et al. 2000). Synthetic monomeric, dimeric and tetrameric poly-APS analogues (Mancini et al. 2004) are currently being investigated for settlement inhibition of *B. amphitrite* cyprids, and some of the analogues show promising antifouling activity (unpubl.).

References

Aizenberg J, Sundar VC, Yablon AD, Weaver JC, Chen G (2004) Biological glass fibers: correlation between optical and structural properties. Proc Natl Acad Sci USA 101:3358–3363

Albrizio S, Ciminiello P, Fattorusso E, Magno S, Pawlik JR (1995) Amphitoxin, a new high molecular weight antifeedant pyridinium salt from the Caribbean sponge *Amphimedon compressa*. J Nat Prod 58:647–652

Almeida AMP, Berlinck RGS (1997) Alcalóides alquilpiridínicos de esponjas marinhas. Quim Nova 20:170–185

Baldwin JE, Spring DR, Atkinson CE, Lee V (1998) Efficient synthesis of the sponge alkaloids cyclostellettamines A-F. Tetrahedron 54:13655–13680

Berlinck RGS, Ogawa CA, Almeida AMP, Sanchez MAA, Malpezzi ELA, Costa LV, Hajdu E, de Freitas JC (1996) Chemical and pharmacological characterization of halitoxin from *Amphimedon viridis* (Porifera) from the southeastern Brazilian coast. Comp Biochem Physiol 115C:155–163

Bunc M, Sepčić K, Turk T, Šuput D (2000) *In vivo* effects of head-to-tail 3-alkylpiridium polymers isolated from the marine sponge *Reniera sarai*. Pflügers Arch 440:R173–R174

Bunc M, Strupi-Šuput J, Vodovnik A, Šuput D (2002a) Toxic effects of head-to-tail 3-alkylpyridinium polymers isolated from the marine sponge *Reniera sarai* in rat. Toxicon 40:843–849

Bunc M, Šarc L, Rozman J, Turk T, Sepčić K, Šuput D (2002b) Intravascular plug formation induced by poly-APS is the principal mechanism of the toxin's lethality in rats/rat tissues. Cell Mol Biol Lett 7:106–108

Calabrese EJ, Baldwin LA (2001) Hormesis: U-shaped dose responses and their centrality in toxicology. Trends Pharmacol Sci 22:285–291

Carroll AR, Scheuer PJ (1990) Four beta-alkylpyridines from a sponge. Tetrahedron 46:6637–6644

Cha JN, Shimizu K, Zhou Y, Christiansen SC, Chmelka BF, Stucky GD, Morse DE (1999) Silicatein filaments and subunits from a marine sponge direct the polymerization of silica and silicones *in vitro*. Proc Natl Acad Sci USA 96:361–365

Clare AS, Freet RK, McClary M Jr (1994) On the antennular secretion of the cyprid of *Balanus amphitrite*, and its role as a settlement pheromone. J Mar Biol Assoc UK 74:243–250

Davies-Coleman MT, Faulkner DJ, Dubowchik GM, Roth GP, Polson C, Fairchild C (1993) A new EGF-active polymeric pyridinium alkaloid from the sponge *Callyspongia fibrosa*. J Org Chem 58:5925–5930

De Vendittis E, Palumbo G, Parlato G, Bocchini V (1981) A fluorimetric method for the estimation of the critical micelle concentration of surfactants. Anal Biochem 115: 278–286

Faimali M, Sepčić K, Turk T, Geraci S (2003a) Non-toxic antifouling activity of polymeric 3-alkylpyridium salts from the Mediterranean sponge *Reniera sarai* (Pulitzer-Finali). Biofouling 19:47–56

Faimali M, Falugi C, Gallus L, Piazza V, Tagliafierro G (2003b) Involvement of acetyl choline in settlement of *Balanus amphitrite*. Biofouling 19 [Suppl]:213–220

Faulkner DJ (2000) Marine pharmacology. Anton Leeuwenhoek Int J Gen M 77:135–145

Fusetani N (2004) Biofouling and antifouling. Nat Prod Rep 21:94–104

Fusetani N, Yasumuro K, Matsunaga S, Hirota H (1989) Haliclamines A and B, cytotoxic macrocyclic alkaloids from a sponge of the genus *Haliclona*. Tetrahedron Lett 30:6891–6894

Fusetani N, Asai N, Matsunaga S, Honda K, Yasumuro K (1994) Cyclostellettamines A – F, pyridine alkaloids which inhibit binding of methyl quinuclidinyl benzilate (QNB) to muscarinic acetylcholine receptors, from the marine sponge, *Stelletta maxima*. Tetrahedron Lett 35:3967–3970

Garaventa F, Faimali M, Sepčić K, Geraci S (2003) Laboratory analysis of antimicrofouling activity of Poly-APS extracted from *Reniera sarai* (Porifera: Demospongiae). Biol Mar Mediterr 10:565–567

Gil L, Gateau-Olesker A, Wong YS, Chernatova L, Marazano C, Das BC (1995) Synthesis of macrocyclic or linear pyridinium oligomers from 3-substituted pyridines. Model synthetic studies toward macrocyclic marine alkaloids. Tetrahedron Lett 36:2059–2062

Goldberg S, Konis Y, Rosenberg M (1990) Effects of cetylpyridinium chloride on microbial adhesion to hexadecane and polystyrene. Appl Environ Microbiol 56:1678–1682.

Hirano K, Kubota T, Tsuda M, Mikami Y, Kobayashi J (2000) Pyrinodemins B-D, potent cytotoxic *bis*-pyridine alkaloids from marine sponge *Amphimedon sp.* Chem Pharm Bull 48:974–977

Kaiser A, Billot X, Gateau-Olesker A, Marazano C, Das BC (1998) Selective entry to the dimeric or oligomeric pyridinium sponge macrocycles via aminopentadienal derivatives. Possible biogenetic relevance with manzamine alkaloids. J Am Chem Soc 120:8026–8034

Kelman D, Kashman Y, Rosenberg E, Ilan M, Ifrach I, Loya Y (2001) Antimicrobial activity of the reef sponge *Amphimedon viridis* from the Red Sea: evidence for selective toxicity. Aquat Microb Ecol 24:9–16

Kobayashi J, Murayama T, Ohizumi Y, Sasaki T, Ohta T, Nozoe S (1989) Theonelladins A – D, novel antineoplastic pyridine alkaloids from the Okinawan marine sponge *Theonella swinhoei*. Tetrahedron Lett 30:4833–4836

Kobayashi J, Murayama T, Kosuge S, Kanda F, Ishibashi M, Kobayashi H, Ohizumi Y, Ohta T, Nozoe S, Sasaki T (1990) Niphatesines A-D, new antineoplastic pyridine alkaloids from the Okinawan marine sponge *Niphates* sp. J Chem Soc Perkin Trans 1:3301–3303

Kobayashi J, Zeng C, Ishibashi M, Shigemori H, Sasaki T, Mikami Y (1992) Niphatesines E – H, new pyridine alkaloids from the Oikinawan marine sponge *Niphates* sp. J Chem Soc Perkin Trans 1:1291–1294

Malovrh P, Sepčić K, Turk T, Maček P (1999) Characterization of hemolytic activity of 3-alkylpyridinium polymers from the marine sponge *Reniera sarai*. Comp Biochem Physiol 124C:221–226

Mancini I, Sicurelli A, Guella G, Turk T, Maček P, Sepčić K (2004) Synthesis and bioactivity of linear oligomers related to polymeric alkylpyridinium metabolites from the Mediterranean sponge *Reniera sarai*. Org Biomol Chem 2:1368–1375

Matsunaga S, Shinoda K, Fusetani N (1993) Cribrochalinamine oxides A and B, antifungal β-substituted pyridines with an azomethine *N*-oxide from a marine sponge *Cribrochalina sp*. Tetrahedron Lett 34:5953–5954

McClelland D, Evans RM, Abidin I, Sharma S, Choudry FZ, Jaspars M, Sepčić K, Scott RH (2003) Irreversible and reversible pore formation by polymeric alkylpyridinium salts (poly-APS) from the sponge *Reniera sarai*. Br J Pharmacol 139:1399–1408

Oku N, Nagai K, Shindoh N, Terada Y, van Soest RWM, Matsunaga S, Fusetani N (2004) Three new cyclostellettamines, which inhibit histone deacetylase, from a marine sponge of the genus *Xestospongia*. Bioorg Med Chem Lett 14:2617–2620

Quiñoà E, Crews P (1987) Niphatynes, metoxylamine pyridines from the marine sponge, *Niphates* sp. Tetrahedron Lett 28:2467–2468

Rao AVR, Reddy GR (1993) First total synthesis of niphatesines A-D and assignment of absolute configuration. Tetrahedron Lett 34:8329–8332

Răşanu N, Barbeş L, Anofriesei F (2000) Derivati de la 3-carboxamidopiridiniu cu actiune antifouling. Rev Chim 51:805–806

Rittschof D (2000) Natural product antifoulants: one perspective on the challenges related to coatings development. Biofouling 15:119–127

Sakemi S, Totton LE, Sun HH (1990) Xestamines A, B, and C, three new long-chain methoxylamine pyridines from the sponge *Xestospongia wiedenmayeri*. J Nat Prod 53:995–999

Schmitz FJ, Hollenbeak KH, Campbell DC (1978) Marine natural products: halitoxin, toxic complex of several marine sponges of the genus *Haliclona*. J Org Chem 43:3916–3922

Scott RH, Whyment AD, Foster A, Gordon KH, Milne BF, Jaspars M (2000) Analysis of the structure and electrophysiological actions of halitoxins: 1,3 alkyl-pyridinium salts from *Callyspongia ridleyi*. J Membrane Biol 176:119–131

Sepčić K (2000) Bioactive alkylpyridinium compounds from marine sponges. J Toxicol Toxin Rev 19:139–160

Sepčić K, Guella G, Mancini I, Pietra F, Dalla Serra M, Menestrina G, Tubbs K, Maček P, Turk T (1997a) Characterization of anticholinesterase-active 3-alkylpyridinium polymers from the marine sponge *Reniera sarai* in aqueous solutions. J Nat Prod 60:991–996

Sepčić K, Batista U, Vacelet J, Maček P, Turk, T (1997b) Biological activities of aqueous extracts from marine sponges and cytotoxic effects of 3-alkylpyridinium polymers from *Reniera sarai*. Comp Biochem Physiol 117C:47–53

Sepčić K, Marcel V, Klaebe A, Turk T, Šuput D, Fournier D (1998) Inhibition of acetylcholinesterase by an alkyl pyridinium polymer from the marine sponge, *Reniera sarai*. Biochem Biophys Acta 1387:217–225

Sepčić K, Poklar N, Vesnaver G, Fournier D, Turk T, Maček P (1999) Interaction of 3-alkylpyridinium polymers from the sea sponge *Reniera sarai* with insect acetylcholinesterase. J Protein Chem 18:251–257

Stierle DB, Faulkner DJ (1991) Antimicrobial *N*-methylpyridinium salts related to the xestamines from the Caribbean sponge *Calyx podatypa*. J Nat Prod 54:1134–1136
Talpir R, Rudi A, Ilan M, Kashman Y (1992) Niphatoxin A and B, two new ichtyo- and cytotoxic tripyridine alkaloids from a marine sponge. Tetrahedron Lett 33:3033–3034
Tsuda M, Hirano K, Kubota T, Kobayashi J (1999) Pyrinodemin A, a cytotoxic pyridine alkaloid with an isoxazolidine moiety from sponge *Amphimedon* sp. Tetrahedron Lett 40:4819–4820
Tsukamoto S, Takahashi M, Matsunaga S, Fusetani N, van Soest RWM (2000) Hachijodines A-G: seven new cytotoxic 3-alkylpyridine alkaloids from two marine sponges of the genera *Xestospongia* and *Amphimedon*. J Nat Prod 63:682–684
Tucker SJ, McClelland D, Jaspars M, Sepčić K, MacEwan DJ, Scott RH (2003) The influence of alkyl pyridinium sponge toxins on membrane properties, cytotoxicity, transfection and protein expression in mammalian cells. Biochim Biophys Acta 1614:171–181
Volk CA, Köck M (2003) Viscosamine: the first naturally occurring trimeric 3-alkyl pyridinium alkaloid. Org Lett 5:3567–3569
Volk CA, Köck M (2004) Viscosaline: new 3-alkyl pyridinium alkaloid from the Arctic sponge *Haliclona viscosa*. Org Biomol Chem 2:1827–1830
Wang GYS, Kuramoto M, Uemura D, Yamada A, Yamaguchi K, Yazawa K (1996) Three novel anti-microfouling nitroalkyl pyridine alkaloids from the Oikinawan marine sponge *Callyspongia sp*. Tetrahedron Lett 37:1813–1816
Whiteley CG, Ngwenya DS (1995) Protein ligand interactions: alkylated pyridinium salts as inhibitors of acetylcholinesterase from *Electrophorus electricus*. Biochem Biophys Res Commun 211:1083–1090

5, 6-Dichloro-1-Methylgramine, a Non-Toxic Antifoulant Derived from a Marine Natural Product

M. Kawamata, K. Kon-ya, W. Miki

Abstract. The laboratory culture of the barnacle, *Balanus amphitrite* has made it possible to supply cypris larvae for antifouling assays all year round. The settlement of cyprids obtained from cultured *B. amphitrite* was indistinguishable from cyprids reared from field-collected barnacles. In laboratory cyprid settlement assays of extracts from marine sessile organisms, antifouling activity was expressed as the 99% inhibitory concentration (IC_{99}), and toxicity as the 30% lethal concentration (LC_{30}). The lipophilic extract of the marine bryozoan, *Zoobotryon pellucidum*, which showed promising antifouling activity, yielded 2,5,6-tribromo-1-methylgramine (TBG) by bioassay-guided isolation. The inhibitory activity of TBG was 6 times as strong as that of bis-(*n*-butyltin)oxide (TBTO), while its toxicity to cypris larvae was one-tenth that of TBTO. A structure-activity relationship study with 155 indole derivatives led to the discovery of the non-toxic antifoulant candidates 5,6-dichlorogramine, 5-chloro-2-methylgramine, and 5,6-dichrolo-1-methylgramine(DCMG), the latter being selected as the antifouling paint ingredient for performance evaluation tests (panel tests) following the results of a preliminary safety tests. A silicone-based antifouling paint containing 5-10% of DCMG was prepared and tested in the field; the painted surfaces remained almost barnacle-free for 1.5 years similar to silicone coatings such as Biox. Since the leaching rate of DCMG from the paint surface could be controlled by the addition of an acrylic acid-styrene copolymer (ASP), the life of the antifouling performance is expected to be improved. Thus, an extremely non-toxic silicone-based antifouling paint containing DCMG is under development.

M. Kawamata
Hydraulic and Bio Engineering Research Section, Civil Engineering Research Institute, Technology Center, Taisei Corporation, 344-1, Nase-cho, Totsuka-ku, Yokohama 245-0051, Japan

K. Kon-ya
Specialty Chemicals R & D Division, Riken Green Co., Ltd., 100 Shibukawa, Shimizu-ku, Shizuoka 424-0053, Japan

W. Miki
Institute for Advanced Technology, Suntory Ltd., 1-1-1 Wakayamadai, Shimamoto, Mishima, Osaka 618-8503, Japan

Progress in Molecular and Subcellular Biology
Subseries Marine Molecular Biotechnology
N. Fusetani, A.S. Clare (Eds.): Antifouling Compounds

1 Introduction

Barnacles, particularly *Balanus amphitrite*, are one of the most problematic macrofouling organisms. The search for antifouling compounds has, therefore, been performed mainly using antifouling assays with cypris larvae obtained from sexually mature *B. amphitrite*. As the spawning season of this barnacle is usually summer, for example, from June to September in Japan, this has seriously limited the period for laboratory antifouling research. However, no attempt had been made to obtain cyprids from laboratory-cultured *B. amphitrite* until we started this project.

It is well known that marine sessile organisms such as sponges, corals and ascidians remain remarkably free from settlement by fouling organisms, thus indicating the presence of defended surface against biofouling (Davis 1991, 1998; Clare 1996; Hadfield 1998; Wieczorek and Todd 1998; Engel and Pawlik 2000). In fact, a number of antifouling compounds have been discovered from marine invertebrates (Fusetani et al. 1996; Miki et al. 1996; Tsukamoto et al. 1996; Fusetani 2004).

In 1989, the Marine Biotechnology Institute started a project to develop a non-toxic antifouling strategy based on the chemical defenses of sessile marine organisms. The major objectives of this project were to: (1) develop a bioassay system to screen for antifouling substances; (2) examine antifouling activity in marine invertebrates; (3) isolate promising antifouling substances and determine their structures; (4) study the structure-activity relationships of promising compounds; (5) evaluate the efficacy and toxicity of the candidate compounds, and (6) develop antifouling paints and to test them in the field.

During the course of this project, we succeeded in obtaining cypris larvae of *B. amphitrite* all year round and found strong antifouling activity in a lipophilic extract of the marine bryozoan, *Zoobotryon pellucidum*. Bioassay-guided isolation afforded 2,5,6-tribromo-1-methylgramine (TBG) from which 5,6-dichloro-1-methylgramine was derived as a non-toxic candidate antifoulant. This compound was incorporated into a prototype coating for field tests.

2 Antifouling Assay

2.1 Laboratory Culture of *B. amphitrite*

The reproduction of *B. amphitrite* is largely confined to the summer in Japan, thereby limiting the time available for bioassays with cypris larvae. We therefore tried to obtain cyprids from laboratory-cultured adults throughout the year. Adult barnacles were collected by immersing acrylic test panels in the ocean for one month in the summer. Between 15 and 20 individuals were selected per panel and fed with *Artemia salina* nauplii in the laboratory. The barnacles began to liberate nauplius larvae in 2 weeks. Cyprids were reared from nauplii according to the method of Kado (1991). Seventy to 80% of nauplii reached the cyprid stage within 6 days.

Five hundred cyprids were transferred in seawater into a box made of transparent acrylic panels, and kept in the dark at 23°C for 5 days. The cyprids (50–70%) settled homogeneously on the inner surface of the acrylic box and metamorphosed into juvenile barnacles. Fifteen to 20 young barnacles were selected by removing unwanted barnacles and reared in a seawater tank with aeration under an 8 h L: 16 h D photoperiod. The diatom, *Chaetocerous calcitrans,* was fed to the barnacles until they reached 3 mm basal diameter, after which they were fed *A. salina*. The barnacles began to liberate nauplii 50–60 days after settlement. To obtain multiple generations, cyprids obtained from 4- to 5-month-old reared barnacles were used. Using this method, the fourth generation of mature barnacles were obtained in 2 years.

2.2 Comparison of the Effect of TBTO on the Settlement Behavior of Cyprids from Reared and Wild Adult Barnacles

We compared the effect of antifouling compounds on the settlement behavior of cyprids from cultured and field-collected adults in order to examine the validity of the assay system. The laboratory-reared cyprids were obtained from 4th-generation laboratory-reared adults. Cyprids from either source were stored in the dark at 5°C to prevent their settlement onto the glass storage beaker (Branscomb and Rittschof 1984; Rittschof et al. 1986) and used for assays within a week.

Samples to be tested for antifouling activity against the cyprids were dissolved in 0.1 mL methanol, applied to a polystyrene Petri dish, allowed to evaporate and then dissolved or suspended in filtered seawater. Ten *B. amphitrite* cyprids were inoculated into each Petri dish. After a 48 h-incubation in the dark, the number settled individuals was counted in order to evaluate the inhibitory activity and toxicity of the test sample compared to the relevant control. In order to confirm the applicability of the assay method using cyprids raised from laboratory-reared adults, the effect of TBTO on their settlement was compared with that of cyprids obtained from field-collected adults.

No significant changes in settlement rates of cyprids from lab-reared adults were observed throughout the year. Settlement rates of the two 'types' of cyprids were similar, and no differences in settlement behavior were observed. Both sources of cyprids died without settlement when exposed to 0.5 to 0.13 ppm TBTO. At a concentration of 0.06 ppm TBTO, the mortality rate fell to 75% and 70% for cyprids reared from laboratory-raised and wild adults, respectively, where individuals of both cyprid 'types' settled onto the bottom of the dishes and metamorphosed into young barnacles. At a concentration of 0.03 ppm, the settlement rates (%) increased as shown in Fig. 1. The IC_{99} values of TBTO for both cyprid types were 0.13 ppm, while the LC_{30} values ranged from 0.04 to 0.05 ppm; no difference was observed in this value between the two types.

Young barnacles raised without *A. salina* in the diet did not grow or release nauplii, suggesting that *A. salina* is indispensable for both growth and maturation of barnacles. On the other hand, wild barnacles feed on a variety of zooplankton and phytoplankton in the hatching season (Barnes 1959).

These results showed that cypris larvae derived from laboratory-reared barnacles could be used to bioassay antifouling substances all year round. The LC_{30} value indicates the harmful effects of a test sample on cyprids, because the mortality rate of the control groups was less than 10%. As shown in Fig. 1, the LC_{30} value was lower than the IC_{99} value in the case of TBTO. All cyprids tested were dead at the IC_{99} values. The data showed that the settlement-inhibitory activity of TBTO toward cyprids is not due to its repellent effect but to its toxicity.

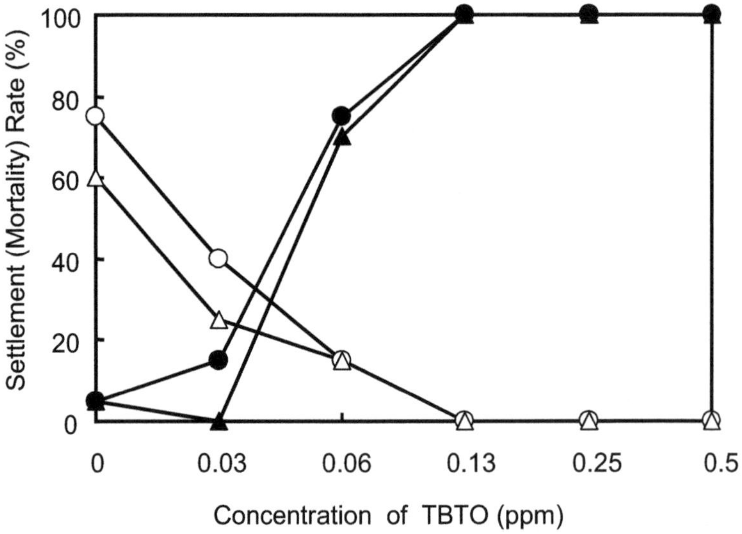

Fig. 1. Response of the "reared" and "wild" cyprids of the barnacle, *B. amphitrite* to various concentrations of TBTO. The settlement rates of "reared" (*open circles*) and "wild" cyprids (*open triangles*) are given in %, and their mortality rates (%) in *closed circles* and *closed triangles*, respectively

In order to search for antifouling compounds that are nontoxic or less toxic than TBTO, it is necessary to obtain biologically active substances not only with repellent effects but also without lethal effects on fouling invertebrates (Ina et al. 1989). In other words, compounds with higher LC_{30} values than IC_{99} values are required.

3
Isolation of 2,5,6-Tribromo-1-Methylgramine from the Marine Bryozoan *Zoobotryon pellucidum*

The acetone extract of *Z. pellucidum*, which showed promising activity in the antifouling assay, afforded the active substance in the yield of 0.006% based on fresh weight by bioassay-guided isolation (Kon-ya and Miki 1994). The active substance was identified as 2,5,6-tribromo-1-methylgramine (TBG); a compound first isolated from the Californian bryozoan, *Z. verticillatum*, as an inhibitor of development of fertilized sea-urchin eggs (Sato and Fenical 1983).

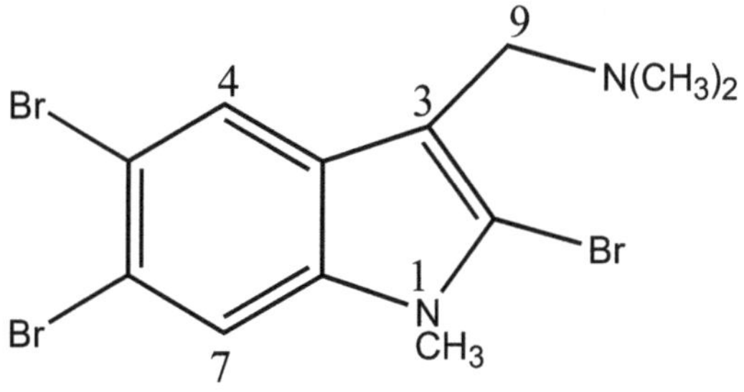

2,5,6-Tribromo-1-methylgramine

TBG showed remarkably potent antifouling activity, while its toxicity toward cyprids was quite low (Fig. 2). At concentrations above 0.03 ppm, all cyprids failed to settle, while they all survived in the concentration range from 0.03 to 0.13 ppm. The mortality rates varied from 7 to 53% in the concentration range 0.25–1.00 ppm. The IC_{99} value of TBG was about one-sixth that of TBTO, whereas its LC_{30} value was only one-tenth of TBTO's (Table 1), thus indicating that TBG inhibits larval settlement by a nontoxic mechanism(s); obviously TBTO exerts its antifouling activity by its toxic property. It should be noted that TBG inhibited the attachment of the blue mussel *Mytilus galloprovincialis* at concentrations as low as 8 ng/cm^2 as

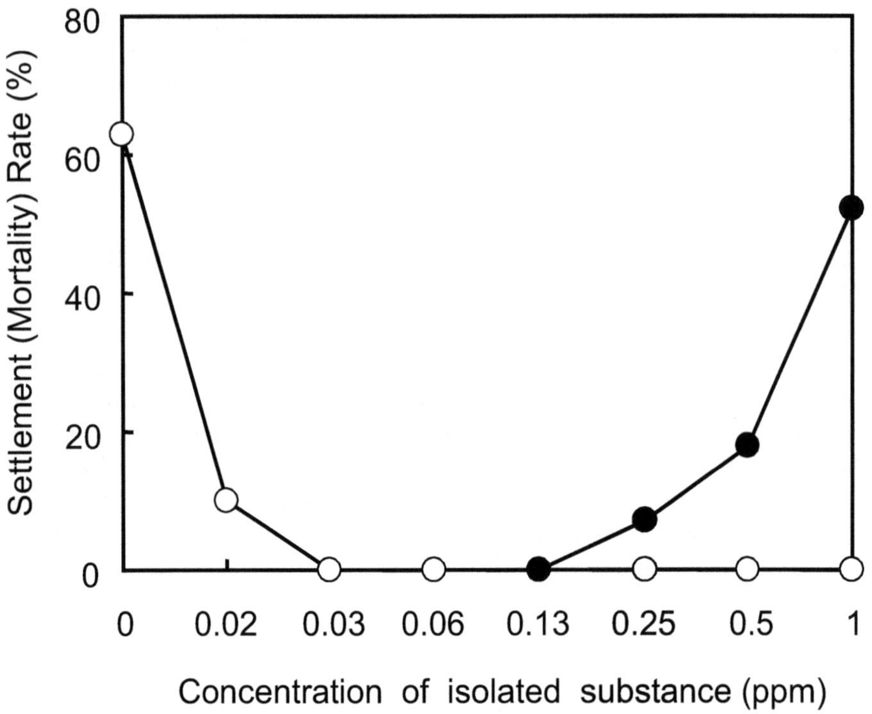

Fig. 2. Antifouling and lethal activities of TBG against *B. amphitrite* cyprids. The settlement (*open circles*) and mortality rates (*closed circles*) of cyprids are plotted against various concentrations of TBG

Table 1. Antifouling (IC_{99}) and lethal (LC_{30}) activities of TBG and TBTO towards *B. amphitrite* cyprids

Substance	IC_{99}(ppm)	LC_{30} (ppm)
TBG	0.03	0.60
TBTO	0.20	0.06

IC_{99}: 99% antifouling activity; LC_{30}: 30% lethal activity

determined by the blue mussel assay (Ina et al. 1989). This activity was 6.8–13.6 times stronger than cupric sulphate. Therefore, TBG was presumed to be a good candidate "non-toxic antifoulant."

4 Structure-Activity Relationships

To discover more effective antifoulants than TBG, a total of 155 gramines and related compounds, which were either synthesized or purchased from

Table 2. Structure-activity relationships of gramine derivatives

Compound								Antifouling activity and lethality (ppm)	
No	R1	R2	R3	R4	R5	R6	R7	IC_{99}	LC_{30}
1	CH_3	Br	$CH_2N(CH_3)_2$	H	Br	Br	H	0.06	1.00<
2	H	H	$CH_2N(CH_3)_2$	H	Br	H	H	0.25	1.00<
3	H	CH_3	$CH_2N(CH_3)_2$	H	H	H	H	0.25	1.00<
4	H	H	H_2CHN- 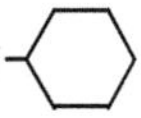	H	H	H	H	0.50	0.72
5	H	H	$CH_2N(CH_2CH=CH_2)_2$	H	H	H	H	0.50	0.38
6	H	H	COPh	H	H	H	H	1.00	1.00<
7	H	CH_3	CO(Ph-4-Cl)	H	H	H	H	1.00	1.00<
8	H	H	Ph	H	H	H	H	–	–
9	H	CH_3	CH_3	H	H	H	H	1.00	0.90
10	H	Ph	H	H	Cl	H	H	1.00	0.64
11	H	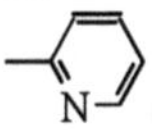	H	H	H	H	H	1.00	1.00<
12	H	CH_3	H_2C- 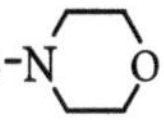	H	H	H	H	0.13	0.65
13	H	H	$H_2C\text{-}N(CH_3)CH_2Ph$	H	H	H	H	0.50	0.64
14	H	H	$CH_2N(CH_3)_2$	H	Cl	Cl	H	0.01	1.00
15	H	CH_3	$CH_2N(CH_3)_2$	H	Cl	H	H	0.01	1.00
16	CH_3	H	$CH_2N(CH_3)_2$	H	Cl	Cl	H	0.06	0.60
TBTO								0.13	0.06

IC_{99}: 99% inhibitory concentration, LC_{30}: 30% lethal concentration
No.14:5,6-dichlorogramine, No.15:5-chloro-2-methylgramine, No.16:DCMG

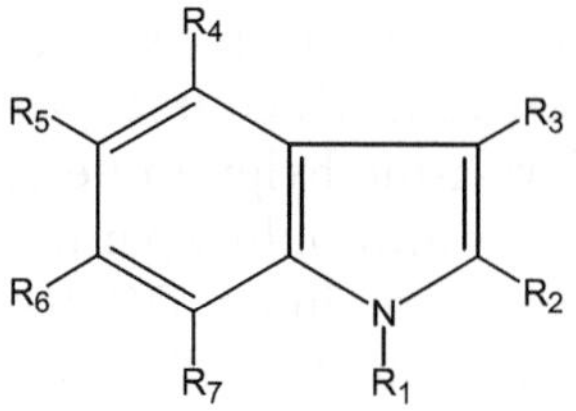

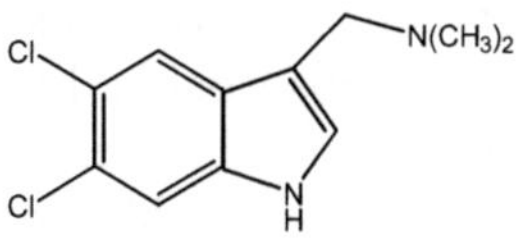

5,6-dichlorogramine

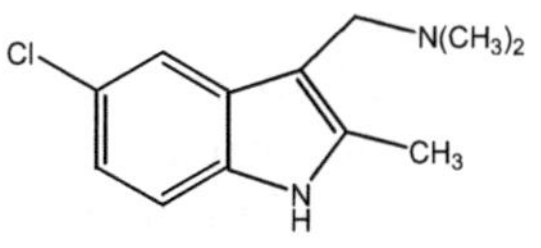

5-chloro-2-methylgramine

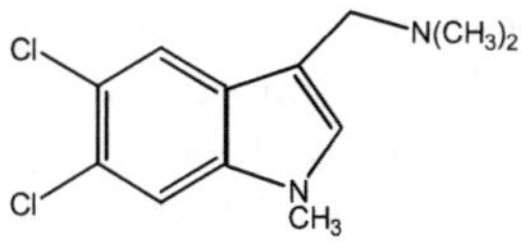

5,6-dichrolo-1-methylgramine

chemical companies, were evaluated for antifouling and toxic activities by our antifouling assay.

Gramine, which is an indole containing a dimethylaminomethyl substituent at the 3-position, showed 79% inhibition of larval settlement at 1 ppm. Further substitution with Br and CH_3 at other positions on the indole skeleton, e.g., 5-bromogramine, 2-methylgramine, and TBG, signify-cantly enhanced the antifouling activity (Table 2). The low LC_{30} values again suggested that their antifouling activity was not due to their toxicity.

It was also found that indole derivatives substituted with a benzoyl group at the 3-position were highly active, but weakly toxic to cypris larvae. The presence of an aminomethyl substituent at the 3-position markedly enhanced activity. Similarly, substitution with a CH_3 group at the 2-position of gramine strongly enhanced settlement-inhibitory activity. These results suggest that the tribromo substituent of TBG is not essential for activity, although it does enhance activity.

Substitution on the indole skeleton with an alkyl, an alkoxyl, a halogen, a carboxyl, a carbonyl, or an acetoxyl moiety, did not enhance the activity, except for 2,3-dimetylindole.

In consideration of these results, 5,6-dichlorogramine (No. 14 in Table 2), 5-chloro-2-methylgramine (No. 15), and 5,6-dichloro-1-methylgramine (No.16) were selected as candidates for further development of a "non-toxic antifoulant."

5
Production of Antifouling Paints

The three candidates above were subjected to a number of tests, such as antifouling activity, stability both in seawater and various organic solvents, and various safety tests. Of course, production cost was a critical factor. Finally, 5,6-dichloro-1-methylgramine (DCMG) was selected as a candidate to develop the new antifouling paint.

In designing the paints containing DCMG (Baier and Meyer 1994; Bakus et al. 1994; Bultman and Griffith 1994; Lindner 1994; Candries 2000), we expected that silicone would have helped DCMG leach gradually from the coating surface at the minimum effective quantity. However, the hydrophilic nature of DCMG prevented us from using silicone alone. Instead, we had to synthesize water-absorbable block polymers (ABPs) that add the water-absorbability to silicone. The addition of ABPs, however, resulted in the rapid leaching of DCMG from the paint surface, which is not favorable for antifouling paints. After several trials, we overcame this problem by adding a suitable quantity of an acrylic acid-styrene copolymer (ASP) to the silicone as described below.

6
Performance Evaluation Tests (Panel Tests)

Steel plates were used for panel tests. The plates were coated as follows: (1) a tar epoxy coating (used for coating ships' hulls) was painted on first as an anticorrosive layer; (2) an epoxy coating formed the second layer; (3) a primer coating was applied as the third layer; (4) the silicone formed the fourth layer, and (5) the silicone containing ABP and DCMG was applied as the fifth and final layer (Fig. 3).

Control panel No. 1 was made of the tar epoxy paint only. Test panel No. 2 was coated with the tar epoxy paint containing 10% DCMG. Test panel No. 3 was coated with Biox, a non-toxic silicone-based coating that has been applied to the cooling system of power plants. Test panels Nos. 4–9 were coated with the test paints altering composition and concentration of DCMG and ABP in the base RTV (room temperature vulcanized) silicone. All panels, fixed on two stainless-steel frames, were placed by ropes at the depth of approximately 2 m in the ocean. Each set of nine panels was tested at seven stations along the Japanese coastline for about 1.5 years.

Figure 4 shows the panels tested for 4 months near Shimizu, central Japan. The surfaces of panels 1 and 2 were completely covered with marine invertebrates, mainly barnacles, ascidians and mussels. As the

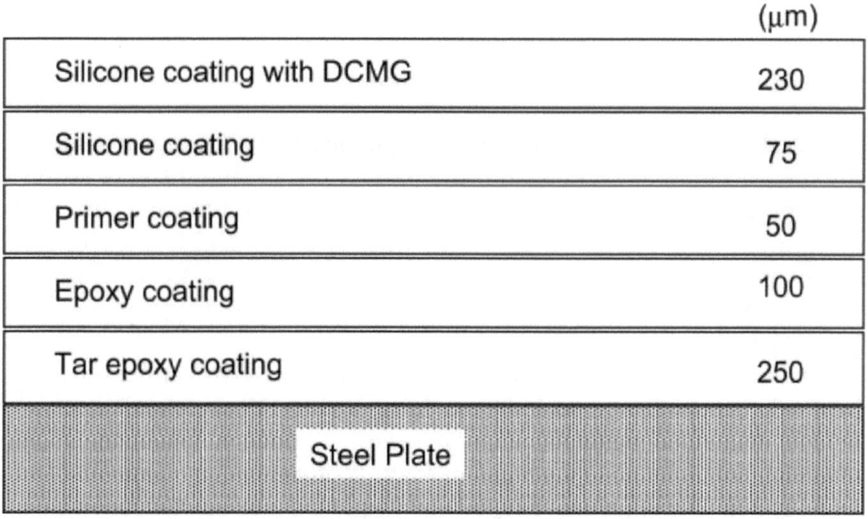

Fig. 3. A typical coating for panel tests, tar epoxy coating: under coating for anticorrosive, epoxy coating: middle coating, primer coating: primer, silicone coating: final coating without additive, silicone coating with DCMG: final coating

Fig. 4. The nine test panels after being immersed for 4 months off Shimizu, central Japan (failure example). No.1, tar epoxy; No. 2, 10% DCMG in tar epoxy; No. 3, BIOX; No. 4, Silicone RTV/ABP1; No. 5, 10% DCMG in silicone RTV/ABP1; No. 6, Silicone RTV/ABP2; No. 7, 5% DCMG in silicone RTV/ABP2, No. 8, 10% DCMG in silicone RTV/ABP2, No. 9, 5% DCMG/3% paraffin in silicone RTV/ABP2

panel 2 coating contained 10% DCMG, this result was unexpected. Perhaps DCMG could not leach from the tar epoxy. In contrast, only slime, a small number of sponges and small tubeworms were observed on panels 3–9. However, no significant difference in antifouling effect was observed between DCMG and the silicone itself.

7 Development of an Effective Antifouling Paint

7.1 Duration of Antifouling Performance

To develop DCMG-containing paints, it is necessary to measure not only the leaching rate of DCMG from the paint's surface but also the retained amount of DCMG in the paint, from which the antifouling life of the antifoulant is estimated. DCMG was quantified by HPLC. Based on the behavior of DCMG in immersed ABP-containing paints, the leaching rate was found to be controllable for practical use (Clare et al. 1995; Clare 1996; Burgess et al. 2003).

The relationship between the leaching rate of DCMG and exposure is inversely proportional; both the leaching rate and the retained amount of DCMG decreased gradually with increasing exposure. When the DCMG leaching rate was above 0.1 $\mu g/cm^2/day$, biofouling was controllable by DCMG. Below 0.01 $\mu g/cm^2/day$, biofouling could not be controlled by the sole use of DCMG. If large amounts of DCMG leach in a short period, the antifouling effect disappears rapidly. Therefore, to develop a long-lived antifouling paint it is necessary to control DCMG-leaching at a minimum effective concentration.

7.2 Control of DCMG-Release and Demonstration Tests

The composition of the test paint was modified to attempt to control the leaching rate of DCMG. Substitution of ABP with the affinity copolymer greatly improved the leaching rate of DCMG from the paint surface and the duration of antifouling efficacy. Perhaps this is a synergistic effect of DCMG and the silicone coating which possesses some antifouling performance by virtue of its low surface energy. Among the affinity copolymers, the acrylic acid-styrene copolymer (ASP) was the most effective in controlling the leaching rate of DCMG.

Demonstration tests were conducted at Kamigoto Island, southern Japan. Test panel No.1 (30×30 cm), which was painted with the tar epoxy coating, was only fouled by barnacles (Fig. 5). Test panel No. 2 coated with Biox, a commercial silicone coating developed by Kansai Paint Co., Ltd., showed good results, although some bryozoans were observed. Panel Nos. 5 and 6 were coated with a silicone Biox-based antifouling paint containing 10% DCMG, and 16% and 20% ASP in coatings 5 and 6, respectively. Neither coating was fouled (Fig. 6). In this experiment, Biox

Fig. 5. Results of panel tests conducted in Kamigoto. Test panel No.1, tar epoxy coating; No. 2, silicone (BIOX) coating. *Circle graphs* show amounts of fouled organisms

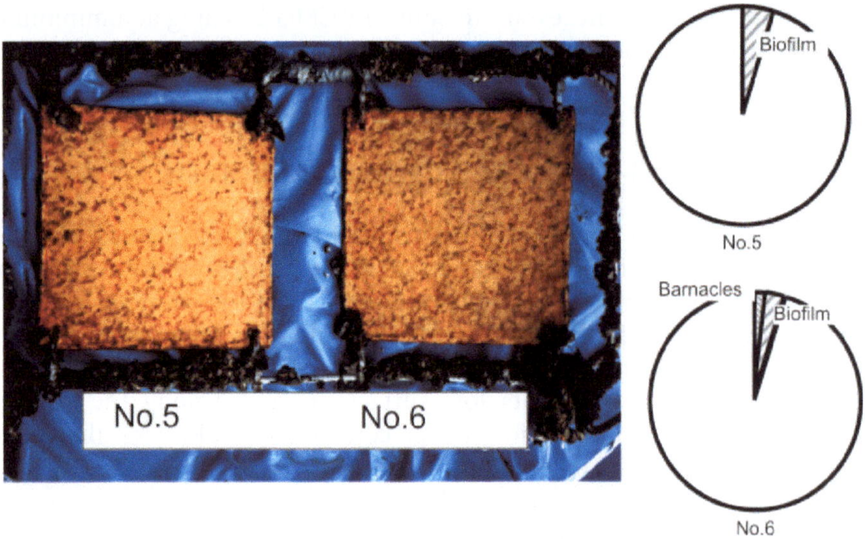

Fig. 6. Results of panel tests conducted in Kamigoto (success example). Test panel No. 5, 10% DCMG in silicone/ASP (5:0.8); No. 6, 10% DCMG in silicone/ASP (5:1). *Circle graphs* show amounts of fouled organisms

was used as base silicone instead of silicone RTV, because Biox is not only much cheaper than silicone RTV but also readily available.

Since ASP seemed to be able to reduce the leaching rate of DCMG from the paint surface as revealed by panel Nos. 5 and 6, various ratios of ASP and silicone (Biox) were prepared for a large-scale experiment. A demonstration test was initiated in 2002 by applying the developed antifouling paint (coating size: 3×6m) onto floating steel oil-fences surrounding oil storage vessels (Kamigoto Oil Storage Co., Ltd.) in Kamigoto Island, from which important information will be collected during the 5-year experimental period.

8 Public Acceptance (Risk Management)

8.1 Safety Test

Various safety tests on DCMG, including acute oral toxicity test, chromosomal aberration test, mutagenic test, Ames test, 28-day repeated dose oral toxicity tests, skin irritation test, bioaccumulation and degradability, were carried out according to the Japanese Chemical

Substances Control Law. The tests were also conducted by such organizations as the Chemicals Evaluation and Research Institute in Japan. DCMG has so far cleared all these tests. Although DCMG was degraded by microbes in activated sludge, DCMG-degrading bacteria were separated from seawater collected from areas where panel tests of DCMG were carried out. DCMG has a half-life of 39 h in sterile artificial seawater as determined by direct photo-degradation. As a result of the tests described above, DCMG was registered as a Designated Chemical Substances with low bioaccumulation but low biodegradability and suspicion of chronic toxicity.

8.2 Risk Evaluation

Ecotoxicity testing of DCMG is now under way using aquatic species such as fish, daphnids and algae according to the Organization for Economic Co-operation and Development test guidelines. As shown in Table 3, DCMG was far less toxic to crustaceans than TBT and TBTO (U'Ren 1983; Meador 1986; Goodman et al. 1998). We have also conducted the risk evaluation using DCMG diffusion simulation, which indicated that the environmental risk of DCMG is low.

9 Summary and Future Perspectives

Using an antifouling assay with cypris larvae from laboratory-reared *B. amphitrite* we discovered DCMG, an extremely effective, non-toxic antifouling agent based on a marine natural product. DCMG was

Table 3. Toxicity of DCMG toward marine and freshwater crustaceans

Crustacean	DCMG	TBTCl
Daphnia magna clone A (freshwater species)	EC_{50} 2.64 mg/mL (24 h) EC_{50} 0.901 mg/mL (48 h) NOEC: 0.0970 mg/mL	LC_{50} 0.0035–0.0060 mg/mL (96 h)[1)]
Americamysis bahia (marine mysid shrimp)	LC_{50} 1.88 mg/mL (48 h) LC_{50} 1.39 mg/mL (96 h) NOEC: 0.126 mg/mL	LC_{50} 0.0011 mg/mL (96 h) (*Mysidopsis bahia*)[2)]
Acartia tonsa (marine copepod)	– –	LC_{50} 0.001 mg/mL (96 h) (in terms of TBTO)[3)] EC_{50} 0.0004 mg/mL (144 h) (in terms of TBTO)[3)]

NOEC: No observed effect concentration; TBTCl: tributyltin chloride, TBTO: bis(tributyltin)oxide
[1)]Meador (1986), [2)]Goodman et al. (1998), [3)]U'Ren (1983)

incorporated into a silicone-based coating and tested in the field with very promising results. It is apparent that the major problem for development of antifouling paints containing DCMG is to control the leaching rate of DCMG from the paint. A 5-year demonstration test has begun, and we look forward to its results. Perhaps a better formulation for DCMG will be developed and safety tests passed.

Acknowledgements: The first half of this work was carried out as part of the Industrial Science and Technology Frontier Program supported by the New Energy and Industrial Technology Development Organization of Japan. The latter half of this work was supported by JOGMEC (Japan Oil, Gas and Metals National Corporation) and ANERI (Advanced Nuclear Equipment Research Institute). We are indebted to the staff of the Marine Biotechnology Institute, Ihara Chemical Industry Co., Ltd., Kansai Paint Co. Ltd., and Taisei Co. Ltd. for their assistance in our research.

References

Baier RE, Meyer AE (1994) Surface analysis of fouling-resistant marine coatings. In: Thompson M-F, Nagabhushanam R, Sarojini R, Fingerman M (eds) Recent developments in biofouling control. Balkema, Rotterdam, pp 286–303

Barnes H (1959) Stomach contents and microfeeding of some common cirripedes. Can J Zool 37:231–236

Bakus GJ, Wright M, Khan AK, Ormsby B, Gulko DA, Licuanan W, Carriazo E, Ortiz A, Chan DB, Lorenzana D, Huxley MP (1994) Experiments seeking marine natural antifouling compounds. In: Thompson M-F, Nagabhushanam R, Sarojini R, Fingerman M (eds) Recent developments in biofouling control. Balkema, Rotterdam, pp 373–381

Branscomb ES, Rittschof D (1984) An investigation of low-frequency sound waves as a means of inhibiting barnacles settlement. J Exp Mar Biol Ecol 79:149–154

Bultman JD, Griffith JR(1994) Fluoropolymer and silicone fouling-release coatings. In: Thompson M-F, Nagabhushanam R, Sarojini R, Fingerman M (eds) Recent developments in biofouling control. Balkema, Rotterdam, pp 383–389

Burgess JG, Boyd K, Armstrong E, Jiang Z, Yan L, Berggren M, May U, Pisacana T, Granmo A, Adams D (2003) The development of a marine natural product-based antifouling paint. Biofouling 19 [Suppl]:197–205

Candries M (2000) Paint systems for the marine industry. Notes to complement the external seminar on antifouling. Department of Marine Technology, University of Newcastle-upon-Tyne, pp 1–27

Clare AS (1996) Marine natural product antifoulants: status and potential. Biofouling 9:211–229

Clare AS, Rittschof D, Price RR, Gerhart DJ (1995) Khellin, a natural product analogue with antifouling activity: laboratory and field studies. In: Bousher A, Edyvean RGJ (eds) Biodeterioration and biodegradation 9. Institute of Chemical Engineers, Rugby, pp 573–580

Davis AR (1991) Alkaloids and ascidian defense: evidence for the ecological role of natural products from *Eudistoma olivaceum*. Mar Biol 111:375–379

Davis AR (1998) Antifouling defence in a subtidal guild of temperate zone encrusting invertebrates. Biofouling 12:305–320

Engel S, Pawlik RP (2000) Allelopathic activities of sponge extracts. Mar Ecol Prog Ser 207:273–281

Fusetani N (2004) Biofouling and antifouling. Nat Prod Rep 21:94–104

Fusetani N, Hirota H, Okino T, Tomono Y, Yoshimura E (1996) Antifouling activity of isocyanoterpenoids and related compounds isolated from a marine sponge and nudibranchs. J Nat Toxins 5:249–259

Goodman LR, Cripe GM, Moody PH, Halsell DG (1998) Acute toxicity of malathion, tetrabromobisphenol-A, and tributyltin chloride to mysids (*Mysidopsis bahia*) of three ages. Bull Environ Contam Toxicol 41:746–753

Ina K, Takasawa R, Yagi A, Yamashita N, Etoh H, Sakata K (1989) An improved assay method for antifouling substances using the blue mussel *Mytilus edulis*. Agric Biol Chem 53:3319–3321

Hadfield MG (1998) Research on settlement and metamorphosis of marine invertebrate larvae: past, present and future. Biofouling 12:9–29

Kado R (1991) Effect of light on the larval development of *Balanus amphitrite* Darwin (Cirripedia). Nippon Suisan Gakkaishi 57:1821–1825

Kon-ya K, Miki W (1994) All-seasonal assay for antifouling substances using reared barnacle larvae. J Mar Biotechnol 1:193–195

Lindner E (1994) Low surface free energy fouling resistant coatings. In: Thompson M-F, Nagabhushanam R, Sarojini R, Fingerman M (eds) Recent developments in biofouling control. Balkema, Rotterdam, pp 305–311

Meador JP (1986) An analysis of photo-behavior of *Daphnia magna* exposed to tributyltin. Proceedings of the organotin symposium ocean 1986 conference. Institute of Electrical and Electronics Engineers, New York, pp 1213–1218

Miki W, Kon-ya K, Mizobuchi S (1996) Biofouling and marine biotechnology: new antifoulants from marine invertebrates. J Mar Biotechnol 4:117–120

Rittschof D, Hooper IR, Costlow JD (1986) Barnacle settlement inhibitors from sea pansies, *Renilla reniformis*. Bull Mar Sci 39:376–382

Sato A, Fenical W (1983) Gramine-derived bromo-alkaloids from the marine bryozoan *Zoobotryon verticillatum*. Tetrahedron Lett 24:481–484

Tsukamoto S, Kato H, Hirota H, Fusetani N (1996) Ceratinamine: an unprecedented antifouling cyanoformamide from the marine sponge *Pseudoceratina purpures*. J Org Chem 61:2936–2937

U'Ren SC (1983) Acute toxicity of bis(tributyltin) oxide to a marine copepod. Mar Pollut Bull 14:303–306

Wieczorek SK, Todd CD (1998) Inhibition and facilitation of settlement of epifaunal marine invertebrate larvae by microbial biofilm cues. Biofouling 12:81–118

Biofilms

J.A. Callow, M.E. Callow

Abstract. Biofilms of bacteria, frequently in association with algae, protozoa and fungi, are found on all submerged structures in the marine environment. Although it is likely that for the majority of organisms a biofilmed surface is not a pre-requisite for settlement, in practice, colonization by spores and larvae of fouling organisms almost always takes place via a biofilmed surface. Therefore, the properties of the latter may be expected to influence colonization, positively or negatively. Biofilms are responsible for a range of surface-associated and diffusible signals, which may moderate the settling behaviour of cells, spores and larvae. However, there is no consensus view regarding either cause and effect or the mechanism(s) by which biofilms moderate settlement. Studies with mixed biofilms, especially field experiments, are difficult to interpret because of the conflicting signals produced by different members of the biofilm community as well as their spatial organisation. Molecular techniques highlight the deficiencies of culture methods in identifying biofilm bacteria; hence, the strains with the most impact on settlement of spores and larvae may not yet have been isolated and cultured. Furthermore, secondary products isolated from cultured organisms may not reflect the situation that pertains in nature. The evidence that bacterial quorum sensing signal molecules stimulate settlement of spores of the green macroalga, *Ulva,* is discussed in some detail. New molecular and analytical tools should provide the opportunity to improve our fundamental understanding of the interactions between fouling organisms and biofilms, which in turn may inform novel strategies to control biofouling.

1 Introduction

Biofilms are formed rapidly on all clean surfaces immersed in the marine environment. The adsorption of organic molecules, which form a conditioning layer on a newly immersed surface (Loeb and Neihof 1975; Baier 1980), alters the physico-chemical condition of that surface (Marshall 1996) and provides a nutrient source for attaching microbial flora. The adsorption of organic molecules to the surface is fast; Little and Zsolnay (1985) observed 0.8 mg m^{-2} of organic matter on a stainless-steel

J.A. Callow, M.E. Callow
School of Biosciences, The University of Birmingham, Birmingham, B15 2TT, UK

Progress in Molecular and Subcellular Biology
Subseries Marine Molecular Biotechnology
N. Fusetani, A.S. Clare (Eds.): Antifouling Compounds

surface after just 15 min of exposure to seawater. Although there is no evidence to suggest that a conditioning film is a prerequisite for microbial attachment, its ubiquitous presence means that the adhesion of primary colonisers including bacteria and algal cells is inevitably influenced by the nature of these films.

Biofilms form through the attachment of bacteria and other single-cell organisms such as algae to a solid surface, followed by growth, cell division and continuing recruitment (Cooksey and Wigglesworth-Cooksey 1995). The species composition of individual biofilms depends primarily on qualitative and quantitative aspects of the inoculum, but light, the substratum, nutrient supply, competition and grazing are all important in determining whether the colonised surface will become a dynamic biofilm ecosystem, or whether higher organisms will become established to produce a climax community (Lock 1981; Biggs 1996). Biofilms on exposed rocky shores are the major food source for grazing gastropods; their grazing activities prevent algal growth, thereby halting succession at the biofilm stage (Hawkins and Hartnoll 1983). However, the importance of the primary colonisers in determining the settlement of higher organisms including macroalgae and invertebrates has only recently been realised. A critical phase during the life cycle of many sessile marine organisms occurs when the planktonic dispersal stages (larvae in the case of invertebrates such as barnacles and hydroids; motile spores in the case of many macroalgae), settle and attach to the substratum. Settlement is a highly selective process involving the exploration of a surface, during which the spore or larva may respond to a range of surface-associated signals (Pawlik 1992; Callow and Callow 2000; M.E. Callow et al. 2002). Many of these signals are likely to originate from the bacterial biofilms that rapidly colonise all surfaces in the marine environment and many studies have demonstrated the influence, both positive and negative, of microbial biofilms and specific strains of bacteria on the settlement and growth of macroalgae (Patel et al. 2003) and invertebrate larvae (Maki et al. 1988; Holmström et al. 1996b; Wieczorek and Todd 1997; Holmström and Kjelleberg 1999a). With increasingly restrictive regulations pertaining to the use of biocidal antifouling agents, environmentally benign methods to control fouling are actively being sought. Consequently, interest in the mechanisms that moderate the colonisation of surfaces by higher organisms has increased, since these may inform new strategies to control biofouling (Wahl 1989; Clare et al. 1992). Many higher organisms have evolved mechanisms that control epibiosis, the most well-known and researched system being the production of halogenated furanones by the red alga *Delisea pulchra* (Steinberg et al. 1997, 1998, 2001, 2002; Steinberg and de Nys 2002). The control of fouling by halogenated furanones and their interference with bacterial signalling pathways is the subject of the chapter by de Nys et al. in this volume.

This chapter will examine the nature of marine biofilms and their impact on colonisation by higher sessile organisms with particular reference to the role of bacterial signalling. The review will briefly cover the formation, composition and physiology of biofilms since knowledge of these processes is fundamental to understanding the role of biofilms in settlement. We will then consider in some detail recent developments in the role of microbial biofilms in the recruitment of the dispersal stages of eukaryotic organisms. The list of examples is not intended to be exhaustive, but aims to illustrate some of the types of signals that might be produced by biofilms as well as some of the reasons for the lack of consensus about the role of biofilms on the settlement of various marine organisms. Particular attention will be given to the green alga *Ulva*, the zoospores of which provide an excellent system in which to investigate details of the signalling processes between biofilms and higher fouling organisms. An important development was the recent demonstration (Joint et al. 2002) that the zoospores of *Ulva* recognize bacterial quorum-sensing acylhomoserine lactones to signal the presence of a biofilm and thus, aid settlement. This was the first report of molecular cross-talk between widely divergent taxa in the marine environment based on a bacterial signalling system. Future research will reveal whether the settling stages of other groups of organisms also respond to bacterial quorum-sensing molecules.

2 Structure and Functional Properties of Marine Biofilms

2.1 Introduction

It is well known that in many ecosystems, the major proportion of microorganisms exist as biofilms. Zobell (1943) first noted that the number of bacteria was markedly higher on surfaces than in the surrounding seawater. During the past two decades, much progress has been made in our understanding of the biofilm habit, especially in industrial and ecological contexts. Major changes in understanding have come about as a consequence of new imaging and molecular technologies. Confocal laser scanning microscopy (CLSM) allowed in situ observation of living hydrated biofilms and showed that the accepted view of biofilms being planar structures was too simplistic: biofilms in reality have a complex 3-D structure made up of variously shaped cell stacks and streamers permeated by channels (Caldwell et al. 1992; Lewandowski 2000). Molecular methods, notably phylotyping of bacteria by 16S rDNA sequencing, shotgun cloning of mixed community DNA, and Fluorescent in situ Hybridisation (FISH) has given an unprecedented ability to characterise biofilm organisms. The use of genomic array methods has allowed studies on global gene regulation as organisms make the transition between planktonic and biofilm states. These and many other aspects of microbial biofilms are amply reviewed in many recent publications which should be consulted for general background information that is applicable to marine biofilms (e.g., Costerton et al. 1995; Lappin-Scott and Costerton 1995; Allison et al. 2000; Evans 2000; Wimpenny 2000; Wimpenny et al. 2000; Doyle 2001; Geesey 2001; Stoodley et al. 2002).

Which particular strains of bacteria colonise a specific surface is the subject of much debate. Mitchell et al. (1996) showed that heterotrophic marine bacteria were capable of swimming at speeds of up to 4 $\mu m\ s^{-1}$, which may imply that colonisation of substrata by bacteria is chemotactically mediated, possibly by gradients of nutrients (reviewed by Amsler and Iken 2001). It is also clear that bacteria are very plastic, having the ability to respond to changing environmental conditions. One of the most profound changes relates to the production of extrapolymeric substances (EPS), which moderate the firm attachment to the substratum and form an intercellular matrix bonding the cells of the biofilm together (Decho and Moriarty 1990). EPS traps and absorbs nutrients, whilst EPS with anionic sites binds metals. The EPS is what the spore or larva 'sees' first when settling on a surface and it has been implicated in moderating settlement, either directly, or indirectly through adsorption of bioactive soluble molecules (see later).

EPS produced by bacterial cells varies greatly in composition. Some polymers are neutral macromolecules but the majority are polyanionic polysaccharides (Sutherland 2001a,b). Proteins, including enzymes and nucleic acids (from lysed cells), are also found in the EPS matrix. Vandevivere and Kirchman (1993) showed that EPS production by attached bacteria was five times higher than by free-living cells. Moreover, there is a qualitative difference between biofilm and planktonic-derived polymers (Allison et al. 1998). In turn, the chemical composition, structure and physical properties of the EPS influence the structure and composition of the biofilm (Hentzer et al. 2001; Nivens et al. 2001). Hydrodynamics also play an important role in determining the structure of biofilms; e.g., *Pseudomonas aeruginosa* forms elongated filamentous downstream 'streamers' under turbulent flow (Stoodley et al. 1999).

Whitely et al. (2001) showed that up- and down-regulation of many bacterial genes occurred during the transition from the planktonic to the biofilm state. For example, within a few minutes of entering the biofilm mode of growth, bacteria repress genes, such as those involved in flagellum synthesis, which are only required during the planktonic phase, and switch on genes to produce EPS to consolidate the structure of biofilm (Davies and Geesey 1995; Garrett et al. 1999). Such observations pose the question of how we can improve our understanding of the biofouling process by studying the interactions between biofilms and settlement since most of our knowledge of the physiology of bacteria and other biofilm organisms has been derived from cells grown in planktonic culture. Understanding the influence of biofilms on the settlement of higher fouling organisms will necessitate a much deeper understanding of the biofilm itself. It is also relevant to note that only a few percent of aquatic microorganisms are culturable, so the strains of microorganism that are most important in moderating settlement of higher organisms may not yet have been identified (Roszak et al. 1984; Colwell et al. 1985; Oliver et al. 1991). A number of studies have shown that the majority of culturable marine bacteria are *Gammaproteobacteria* whilst molecular approaches show *Alphaproteobacteria* to be most abundant (Hagström et al. 2000). In this context, Imhoff and Stohr (2003) reported little or no correspondence between the bacteria associated with sponges identified by culture and molecular techniques.

2.2 Phylogenetic Identification in Complex Microbial Communities

Until recently, identification of microorganisms required the isolation of pure cultures followed by physiological and biochemical tests. Clearly the main disadvantage of this traditional approach is that only culturable microorganisms could be identified. The advent of molecular microbiology has radically altered our understanding of natural microbial communities. Schmidt et al. (1991) used the 16S rRNA approach to identify dominant bacteria in a marine picoplankton community. Shotgun clone libraries prepared from mixed community DNA were screened with specific 16S rDNA probes followed by 16S rDNA sequencing of the selected clones. Giovannoni et al. (1990) used PCR to selectively amplify 16S rRNA gene fragments from mixed template DNA. They reported the presence of a novel microbial group, SAR 11, but none of the SAR 11 sequences were identical to any cultivated marine bacteria. Delong et al. (1993) introduced restriction fragment length polymorphism (RFLP) to identify similar sequences among 16S rDNA clones. RFLPs showed a difference in population structure between free-living and macroaggregate-associated bacteria in the marine environment. Phylogenetic identification revealed that macroaggregate-associated rRNA clones were related to *Cytophaga, Planctomyces* and *Gammaproteobacteria*, while the free-living bacterial population was closely related to *Alphaproteobacteria*, which also reinforced previous findings (Giovannoni et al. 1990; Schmidt et al. 1991). Eilers et al. (2000) combined 16S rDNA cloning with cultivation of North Sea bacterioplankton using different oligotrophic media. The abundance of 16S rDNA clones and cultured isolates was determined using FISH, which showed that none of the readily culturable genera *Pseudoalteromonas, Alteromonas* and *Vibrio* were detected among the 16S rDNA clones, or by FISH. Conversely, SAR 86 clusters and a *Cytophaga-Flavobacteria* cluster were detected by FISH- but not by the cultivation approach.

FISH using 16S rRNA probes was first introduced by Delong et al. (1989) as a method for detecting specific bacteria and their spatial distribution in a complex microbial community. Specificity of these probes can be used to detect taxonomic groups at genus and species levels, as well as higher phylogenetic groups. The technique is now used extensively for the identification and quantification of bacteria in complex microbial communities by targeting specific but unknown and unculturable bacteria (Amann et al. 1995). Cottrell and Kirchman (2000) showed that a marine bacterioplankton community analysed by FISH was substantially different to that determined from clone libraries. On average, only 10% of DAPI-stained bacteria were *Alphaproteobacteria* but 55% were detected by cloning analysis, while the Bacteriodetes (formerly CFB group) was shown to be dominant by FISH with 30% of DAPI-stained bacteria belonging to this group compared with only 10% or fewer clones detected by cloning analysis. In contrast, Glöckner et al. (1999) showed by FISH that *Betaproteobacteria* were dominant in freshwater systems while *Alphaproteobacteria* dominated marine waters.

The techniques of denaturing gradient gel electrophoresis (DGGE) or temperature gradient gel electrophoresis (TGGE), have also been used to profile complex microbial communities including those forming mats and biofilms (Ferris et al. 1996; Ferris and Ward 1997; Muyzer 1999).

2.3 Algal Biofilms

Diatoms are usually considered the most common and abundant of the early eukaryotic colonisers being ubiquitous on submerged surfaces (Jackson and Jones 1988; Callow 2000). Diatoms are unicells or colonial algae, characterised by the presence of an elaborately ornamented silica frustule and chloroplasts containing the pigment fucoxanthin, which masks the chlorophyll, hence their brown colour (Round et al. 1990; van den Hoek et al. 1995). Cells range in size from a few to several hundred micrometres. Pennate diatoms have bilateral symmetry and are the abundant types in benthic habitats, particularly those genera that possess an elongate slit, the raphe, in one or both valves of the frustule. Raphid diatoms may be sessile or motile but in both cases, EPS is secreted via the raphe(s), which in the former provides the means for adhesion whilst in the latter, it provides the mechanism for both adhesion and motility by gliding (Edgar and Pickett-Heaps 1984). Raphid diatoms are probably the most important biofilm algae, being instrumental in the primary colonisation of submerged substrata (Jackson and Jones 1988).

Diatom EPS (Wetherbee et al. 1998) is a multicomponent, mucilagenous, organic bioadhesive complex found exterior to the plasma membrane. In common with the EPS of many bacteria, the major matrix components are acidic polysaccharides that are frequently carboxylated or sulphated (Daniel et al. 1987; Hoagland et al. 1993; Wustman et al. 1997; Chiovitti et al. 2003a,b). Proteoglycans are also implicated in both adhesion and gliding motility (Lind et al. 1997; Wetherbee et al. 1998).

The EPS of some benthic diatoms may be secreted from points in the frustule other than the raphe, e.g., an apical pore field, becoming elaborated into structures such as pads and stalks (Daniel et al. 1987; Hoagland et al. 1993). Stalks and other extracellular structures that elevate the cell body above the substratum confer advantages in competition for nutrients and light and impart a significant topography to the surface that may be of importance in spore and larval settlement.

Over a period of time, the EPS of both bacteria and algae is likely to be modified by secondary adhesion processes such as cross-linking by phenolics (Wustman et al. 1997) and/or the addition of other molecules (Wetherbee et al. 1998).

3 *Ulva* Zoospores – a Model for Studying the Influence of Marine Microbial Biofilms on Biofouling Processes

Ulva (formerly *Enteromorpha* [Hayden et al. 2003]) is a cosmopolitan intertidal macroalga and a major contributor to marine biofouling. It produces motile, quadriflagellate, naked spores (zoospores) that settle through a process involving sensing of a surface, and temporary adhesion (Callow et al. 1997) followed by discharge of a glycoprotein adhesive (Stanley et al. 1999; J.A. Callow et al. 2000, 2002) to form a permanent attachment. Cell wall formation and growth follows with development into a vegetative thallus. Zoospores explore and attach temporarily to a surface, but detach if it is not optimal; i.e., there appears to be sensing of surfaces to select suitable sites before irreversible attachment occurs (Callow et al. 1997; Callow and Callow 2000). Many factors influence the attachment of zoospores including negative phototaxis, thigmotaxis, chemotaxis (Callow and Callow 2000), surface chemistry and wettability (M.E. Callow et al. 2000; Ista et al. 2004), and surface topography (M.E. Callow et al. 2002; Hoipkemeier-Wilson et al. 2004; Granhag et al. 2004). Some or all of these may contribute to the location of a suitable surface on which to settle in the natural environment. Bacterial biofilms may be a source of some of these signals and consideration of this forms the focus of this section of the review.

3.1 Influence of Microbial Biofilms on Zoospore Settlement

Microbial biofilms are present on all submerged surfaces in the marine environment and thus change the properties of those surfaces. Dillon et al. (1989) found enhanced settlement of *Ulva (Enteromorpha)* spores on mixed microbial biofilms. Thomas and Allsopp (1983) showed that the biofilms of some strains ascribed to the *Pseudomonas/Alteromonas* group and *Coryneform*, enhanced the number of *Ulva* (*Enteromorpha*) germlings detectable after incubation with spores while other strains were inhibitory; spore settlement was not explored directly. A wider range of strains was tested with spores of *Ulva lactuca* (Holmström et al. 1996a). Sixteen out of 24 strains tested were inhibitory, three of which were darkly pigmented isolates that inhibited algal spore settlement by lysing spores. These strains showed a close phylogenetic affiliation with *Pseudoalteromonas tunicata* (Egan et al. 2002). In quantitative analyses of spore settlement in relation to biofilms, Joint et al. (2000) demonstrated that while *Ulva (Enteromorpha)* spores settled on control glass slides without a biofilm, the level of settlement was significantly enhanced by biofilm assemblages formed from natural seawater; there being a positive correlation between the number of bacteria in the biofilms and the number of spores that attached. Patel et al. (2003) extended the studies of Joint et al. (2000) by examining the relationship between spore settlement and biofilms of specific strains of bacteria of known identity. Bacteria were isolated from surface biofilms of *Ulva* plants and rock surfaces in close proximity to plants and identified by 16S rDNA sequencing. Biofilms of specific bacterial strains, and of different age, were examined for their ability to influence spore settlement. Results showed that monospecies biofilms of specific strains of bacteria could stimulate spore settlement but that these effects were dependent on the age, and therefore the density of the biofilm – which was corrected for in the analysis. The effect of bacterial biofilms on the settlement of spores could not be assigned to species or even genus level as strains within species showed dissimilar effects. However, within the *Gammaproteobacteria*, the genera *Vibrio* and *Shewanella* contained a greater proportion of more stimulatory strains than *Pseudoalteromonas* and *Cobetia* (syn. *Halomonas*). The Bacteroidetes (previously the Cytophaga-Flavobacteria-Bacteroides (CFB) group) also contained more stimulatory strains than *Pseudoalteromonas* and *Cobetia*. The activities of some of the *Pseudoalteromonas* strains are discussed later. Joint et al. (2000) used image analysis to explore the spatial relationship between bacterial cells in mixed-species biofilms and attached zoospores. Preferential spore settlement on, or in close proximity to some bacterial clumps was observed, suggesting that these microcolonies represent direct cues for spore settlement. Patel et al. (2003) extended this analysis to biofilms of individual strains but the results indicate a more complex situation since several different categories of association can be recognised and overall, the spatial association between spores and bacteria appears to be independent of the overall quantitative influence of bacterial cells on spore settlement. However, the most interesting strains were those that both enhanced spore settlement and showed a close spatial association with bacterial cells (Fig. 1), and this category was represented by eight strains distributed across a number of taxa.

3.2
Recognition of *N*-acylhomoserine Lactones by Zoospores

The non-random patterns of settlement of zoospores on bacterial cells or microcolonies of specific bacterial strains suggested that zoospores of *Ulva* are capable of responding to signals directly produced by these cells.

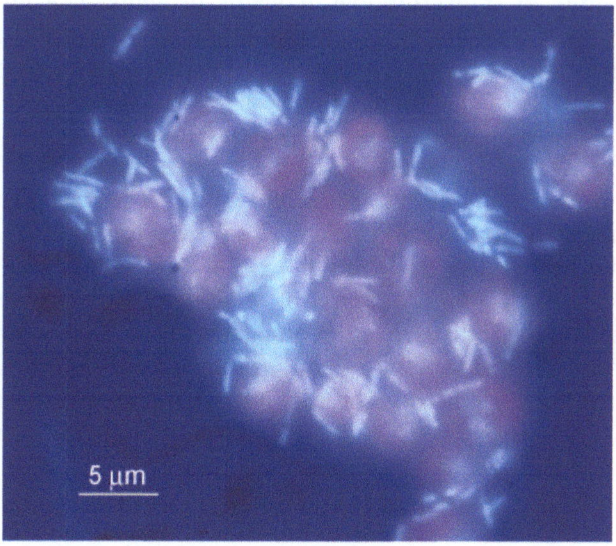

Fig. 1. Spatial association between zoospores of the green alga *Ulva linza* and biofilm bacteria. Zoospores were settled on a monospecific bacterial biofilm formed on a glass coverslip and prestained with the vital stain Calcofluor White ST. After washing to remove unsettled spores, the coverslip was mounted on a glass slide and viewed through the coverslip by UV epifluorescence. The specimen was focussed to bring the bacterial cells, which fluoresce intensely white, into sharp focus; the pink, autofluorescent, chlorophyll-containing spores being below this plane of focus. The image was prepared by Pratixa Patel and James Callow.

The range of potential chemical cues that would be presented by a biofilm is immense but one possibility, first examined by Joint et al. (2002), is that zoospores might use, as recognition cues, those signals that bacteria themselves use to communicate within biofilms in the process of quorum sensing (QS). Quorum sensing is now recognised as a ubiquitous cell-to-cell communication mechanism in bacteria and employs a chemically diverse range of small diffusible signal molecules (Bassler et al. 1994; Miller and Bassler 2001; Swift et al. 2001; Withers et al. 2001; Taga and Bassler 2003). One of the principle types of signal used by many Gram-negative bacteria is the *N*-acyl homoserine lactones (AHLs) (Davies et al. 1998; Huber et al. 2001). These consist of five-membered homoserine lactone rings containing varied amide-linked acyl side chains (Fig. 2). The *N*-acyl moieties of the naturally occurring AHLs identified to date, range from 4 to 14 carbons in length and may be saturated or unsaturated, with or without a substituent (usually hydroxy or oxo) on the carbon at the C3 position of the *N*-linked acyl chain (Withers et al. 2001). AHLs pass out of and into bacterial cells, and as the population of bacteria increases so does the AHL concentration. Once they reach a critical threshold concentration, AHLs act as co-inducers by activating members of the LuxR or LuxN family of transcriptional activator proteins. The LuxR(N)/AHL complex then drives the expression of multiple target genes. At least three types of AHL synthases have been identified which correspond to the LuxI, LuxM and HdtS proteins (Swift et al. 2001).

Joint et al. (2002) used three approaches to determine whether *Ulva* zoospores sense the presence of bacteria by detecting AHLs. The first of these used mutants of *Vibrio anguillarum* defective in AHL production. *V. anguillarum* is a marine bacterium with two linked AHL-dependent quorum sensing regulatory circuits (the VanIR and VanMN circuits) (Milton et al. 1997, 2001). Biofilms of wild type *V. anguillarum* (strain NB10) were very effective in attracting zoospores. Zoospores were also attracted to biofilms formed by a *V. anguillarum vanI* mutant which lacks the synthase required to produce *N*-(3-oxodecanoyl)-L-homoserine lactone (3-oxo-C10-HSL) (Fig. 2), but has normal production of two other AHLs (*N*-hexanoyl-L-homoserine lactone [C6-HSL] and *N*-(3-hydroxyhexanoyl)-L-homoserine lactone [3-hydroxy-C6-HSL]) (Fig. 2). The latter AHLs are synthesized via the *V. anguillarum vanM* gene product. However, no stimulation of attachment was observed with a *vanM* mutant, which does not produce C6-HSL or 3-hydroxy-C6-HSL (Milton et al. 2001). Since the AHLs produced via *vanM* are also required for the production of 3-oxo-C10-HSL, the *V. anguillarum vanM* mutant is deficient for all 3 AHLs. The double *vanIM* mutant also failed to stimulate zoospore attachment. The second approach was to perform

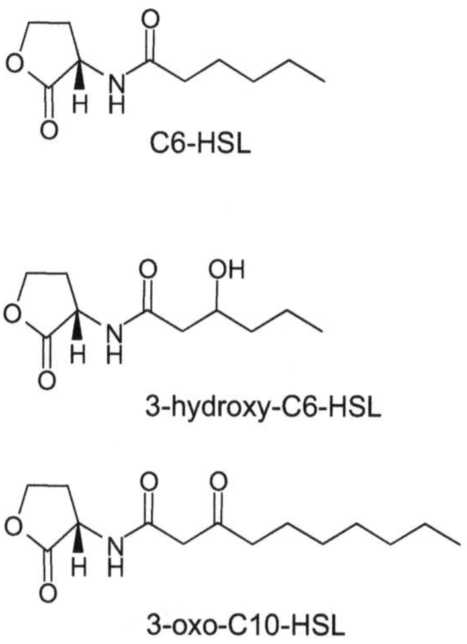

Fig. 2. Structures of acyl-AHLs produced by quorum-sensing *Vibrio anguillarum* bacteria

zoospore settlement assays using *Escherichia coli* strain K12 expressing the recombinant AHL synthase genes *vanI* or *vanM*. *E. coli* K12 strains do not produce AHLs and its biofilms are not attractive to *Ulva* zoospores. Biofilms of the recombinant strains do produce AHLs (Milton et al. 2001) and zoospore settlement was enhanced on biofilms producing 3-oxo-C10-HSL from *E. coli* BL21 containing pVanI (expressing *vanI*), and both C6-HSL and 3-hydroxy-C6-HSL from *E. coli* DH5α containing pDKVanM (expressing *vanM*). The direct involvement of AHLs in zoospore settlement was finally confirmed in experiments using synthetic AHLs presented to zoospores via agarose films. AHLs with acyl chain lengths from C_6 to C_{14} increased spore settlement. Control experiments with ring-opened, non-functional AHLs failed to enhance settlement. The addition of AHLs (C6-HSL, 3-hydroxy-C6-HSL or 3-oxo-C10-HSL) to the seawater covering biofilms of wild-type *V. anguillarum* significantly reduced stimulation of zoospore settlement by this bacterium. These results imply that zoospores are capable of detecting a gradient of AHLs diffusing from the agarose, but if this gradient is eliminated by increasing the concentration of dissolved AHLs in the seawater in which zoospores are swimming, attachment is reduced because chemoattraction is disrupted.

Further experiments on AHL recognition by zoospores were performed by Tait et al. (2005). Biofilms of *V. anguillarum* strains expressing the inducible AHL-degrading lactonase enzyme AiiA, produced no AHLs and were not attractive to zoospores. Biofilms of *V. anguillarum* wild type treated with chloramphenicol or short-term UV exposure to kill the biofilm without disrupting its physical integrity, did not stimulate zoospore settlement, suggesting that zoospores respond to something produced by live biofilms and not a physical feature like topography. Experiments with a Green Fluorescent Protein AHL-reporter strain of *V. anguillarum* showed that zoospores settled over AHL-producing microcolonies. The response of zoospores to different AHL molecules was examined and it was shown that AHLs with longer (>6 carbon) *N*-acyl chains were more effective than shorter chain lengths- although factors such as different rates of diffusion, differential solubility and differential rates of breakdown complicate this conclusion.

3.3 Ecological and Applied Significance of Interspecific AHL Signalling in Complex Marine Communities

Bacterial biofilms have often been suggested as playing an important role in the development of algal communities since a number of studies have demonstrated influences of individual bacterial strains on algae (recently reviewed by Holmström and Kjelleberg 2000). However, most reported effects are inhibitory and until recently there was little information on stimulatory interactions that might influence the recognition and adhesion of algal spores to marine surfaces. The studies on AHLs reported above show that algal attraction to a surface is mediated via QS signalling molecules produced by bacterial biofilms – at least in laboratory experiments. Less clear is what the ecological relevance of this is. One can imagine that detection of AHL QS molecules may play an important role in the ecological success of *Ulva* by aiding the selection of attachment sites – but it is not yet clear what beneficial properties such attachment sites might provide for such a mechanism to be ecologically advantageous. Then there is the question as to whether the sorts of concentrations of AHLs that spores recognise in laboratory experiments (typically in the μM range) are relevant to natural physiological concentrations. This again is difficult to answer because the AHL production by biofilms is in itself dynamic and an additional complication is that AHLs show a short half-life in the alkaline conditions of seawater (Tait et al. 2005). Additional information that would assist our understanding of this type of 'molecular cross-talk' would be to understand the mechanism by which an AHL molecule produces an effect on the spore. For example, does *Ulva* contain and express homologues of recognised bacterial AHL-receptors – or do the AHLs bind to molecules unrelated to known receptors? Are other classes of QS molecule recognised by zoospores or is the effect AHL-specific? What is the mechanism by which zoospore settlement is changed? Is it a true chemotaxis in response to a concentration gradient or is it by some other mechanism such as chemokinesis (Amsler and Iken 2001)? What is becoming clear is that AHLs have biological properties beyond their role in bacterial cell-cell communication. In particular, evidence is accumulating to show that eukaryotes respond to AHLs to assist them in making the appropriate response to either pathogenic or mutualistic bacteria. Certain AHLs and in particular, long chain AHLs are known to possess immunomodulatory (Telford et al. 1998) and pharmacological properties (Lawrence et al. 1999; Gardiner et al. 2001). Mathesius et al. (2003) showed that the legume *Medicago truncatula* responds to nanomolar to micromolar concentrations of bacterial AHLs by expressing specific sets of genes and proteins that may have critical functions in controlling infection. Furthermore, the higher plant itself, and the green alga *Chlamydomonas* (Teplitski et al. 2004) both produce compounds that mimic bacterial signals or interfere with QS processes in bacteria. Elsewhere in this volume, de Nys et al. outline the work on the halogenated furanones of the red alga *Delisea pulchra* and their potential role in controlling natural epibiosis. These furnanones have a number of biological activities, but some of them specifically inhibit AHL-regulated behaviours in Gram-negative bacteria (Givskov et al. 1996) by competing for the AHL receptor (Manefield et al. 1999). The furnanone QS-mimics of *D. pulchra* help the alga to control the colonisation of its surfaces by epibionts.

Despite the many as yet unanswered questions, evidence is gradually emerging to show that higher organisms can 'listen in' on the signals used by biofilm-forming organisms they come into contact with, and this may come to be seen more generally, as the consequence of a significant co-evolutionary process. The question still remains - what possible advantage do *Ulva* zoospores gain by recognising the signals used by some biofilm-forming bacteria as indicating a suitable site for spore settlement and subsequent growth of the plant? What is clear is that *Ulva* zoospores do not need a biofilm to settle - a newly presented pristine surface is quickly colonised by settled zoospores. But in nature a pristine surface never exists - biofilmed surfaces are ubiquitous. The *Ulva* plant itself has a characteristic biofilmed surface and at least some of the bacteria involved may be regarded as episymbiotic since axenic plants of *Ulva* do not grow well and fail to produce a normal morphology. Provasoli and Pintner (1980) showed that *Ulva lactuca* plants developed an atypical 'pincushion' morphology when grown axenically; the plants reverting to a normal 'foliaceous' morphology when inoculated with marine bacteria. Similar effects on morphology after removal of bacteria have been observed for other related green algal species (Tatewaki 1983; Nakanishi et al. 1996, 1999; Matsuo et al. 2003). The most effective strains of bacteria in inducing both axenic *U. pertusa* (Nakanishi et al. 1996) and

M. oxyspermum (Matsuo et al. 2003) to revert to a more typical morphology were in the Bacteriodetes group (previously *Cytophaga-Flavobacterium-Bacteriodetes* (CFB)). It has been speculated elsewhere (Tait et al. 2005) that those biofilm-forming bacteria that stimulate spore settlement could act as important 'founder strains' helping the plant to set up an episymbiotic relationship with those strains that are important for normal growth. However, Marshall et al. (in preparation) found no clear relationship between bacterial strains that induced spore settlement and those that caused normal vegetative development, so at this stage the hypothesis is unproven.

It has often been speculated that QS-signal mimics have application potential in areas where it is desirable to control biofilm formation or function, including biomedical, agricultural and environmental contexts. The recognition of a biofilmed surface by *Ulva* spores, through AHLs, may suggest novel ways of controlling the settlement of spores of this important fouling organism and this is one context in which the furanones of *Delisea pulchra*, or their synthetic equivalents, are currently being explored. It will be interesting to see whether other groups of marine organisms, such as barnacles and hydroids, share with *Ulva* the ability to recognise and respond to bacterial biofilm signals.

4 Interactions of Biofilms and Bacterial Metabolites with Invertebrate Larvae

4.1 Mixed and Single Species Biofilms

As far as is known, no surface after immersion in the sea, including the most potent biocidal antifouling coatings, is devoid of a bacterial biofilm. In most cases, the biofilm will be a mixed microfouling community consisting of bacteria, diatoms and other microalgae, protozoa and fungi. Thus, in both the natural ecosystem and on artificial substratums, settling larvae will encounter a complex biofilmed surface. As early as 1935, it was suggested that marine bacterial biofilms influenced the settlement of invertebrate larvae (Zobell and Allen 1935). The effects of a biofilm may affect larval behaviour in a number of ways: firstly by releasing soluble molecules that moderate chemotaxis or chemokinesis; secondly by affecting the settlement behaviour of the larvae and thirdly by affecting larval adhesion and subsequent metamorphosis. In some cases it may be difficult to distinguish between the different effects. Field observations are more ecologically relevant but shed little light on mechanisms since there is a multitude of competing cues from both prokaryotes and eukaryotes, and only the end effect (settlement induction or inhibition) is measured. A summary of the literature on the effect of biofilm on the settlement and metamorphosis of invertebrate larvae can be found in Railkin (2004) and Wieczorek and Todd (1998).

For many invertebrates such as barnacles, surface-associated cues such as topography (Hills and Thomason 1998), surface polarity (Roberts et al. 1991), biochemical cues from conspecific adults (Thompson et al. 1998) and biofilms (Le Tourneux and Bourget 1988) are well known to be important in settlement (see Railkin 2004). On an artificial substratum especially, larvae may be confronted by surface cues provided by the substratum, as well as surface cues provided by the biofilm. Pawlik (1992) summarised the earlier literature on the role of biofilms on settlement and recruitment of invertebrate larvae; later summaries can be found in chapters by Hadfield and Paul (2001) and the monograph by Railkin (2004). The nature of the biofilm and likely its signalling system may be changed by the properties of the underlying substratum. Dalton et al. (1994) showed dramatic morphological changes of a marine bacterium SW5 when cultured on a hydrophobic compared to a hydrophilic surface. It is likely that many more subtle changes are also expressed because of the properties of the surface. Field observations are also complicated by the patchiness of the biofilm. For example, Hudon et al. (1983) showed barnacle cyprid settlement was confined to areas devoid of detritus and diatoms but a bacterial biofilm was almost certainly present.

Detailed studies using natural biofilmed and clean surfaces revealed that there was generally higher settlement of invertebrate larvae on the biofilmed surfaces, although this was not the case for all organisms or at all sampling times (Todd and Keough 1994; Keough and Raimondi 1995, 1996), or for all biofilm ages (Wieczorek et al. 1995; Wieczorek and Todd 1997, 1998). Many studies show inhibition of settlement of cyprids of *Balanus amphitrite* by bacterial biofilms (Lau and Qian 2000; Olivier et al. 2000).

Laboratory experiments have many limitations but are essential if the mechanisms that moderate settlement are to be revealed. So far, the majority of studies have used multi-species natural biofilms, which may promote or inhibit settlement. Numerous examples are cited in Maki (1999) and Railkin (2004). Some studies have employed single species biofilms but again, the effects of biofilms on larval settlement vary; differences are likely to be due to different methodologies for preparing the biofilms, the use of different bacterial strains, the stability of some of the cultures plus the transitory nature of compounds that signal different phases of biofilm development. For example, *Cobetia marina* (syn. *Halomonas marina, Pseudomonas marina, Deleya marina*) is reported to stimulate settlement of the spirobid polychaete, *Janua brasiliensis*, stimulate or inhibit settlement of *Balanus amphitrite* and inhibit settlement of the bryozoan *Bugula neritina* (reviewed in Maki 1999). Maki and co-workers (Maki et al. 1990; Maki 1999) suggested that the wettability of the substratum influenced whether barnacle cyprid settlement was stimulated or inhibited by *C. marina* because the bacterium expressed genes defining EPS in response to the substratum on which it was cultured, although the 3-D topography of the biofilm may also act as a settlement cue (Tsurumi and Fusetani 1998).

Larval behaviour of *Janua brasiliensis* in response to *Cobetia marina* biofilms was attributed to interactions between the bacterial EPS and larval surface lectins (Kirchman et al. 1982a,b). Strong inhibition of settlement was also found in response to D-glucose whilst biofilms treated with Concanavalin A (specificity for mannose and glucose) no longer inhibited larval settlement. These and other data led Kirchman to propose that settlement was a lectin-mediated event involving recognition and binding between the larval surface and the EPS of the bacterium. A recent study of settlement of *Balanus amphitrite* cyprids on single species biofilms (Khandeparker et al. 2003) also suggests that lectins in the EPS may be involved in recognition. Other studies have also implicated EPS recognition events in larval settlement (Holmström and Kjelleberg 1999a). Lau et al. (2003b) used three strains of bacteria that inhibited settlement of cyprids of *Balanus amphitrite*. The inhibitory effect appeared to necessitate contact with the biofilm since inhibition occurred for both living and dead biofilms. On the other hand, bacterial EPS appeared to facilitate the settlement of *Hydroides elegans* (Lau et al. 2003a). Bacterial exopolymers may also bind soluble bioactive molecules produced by the bacteria, thereby rendering the EPS 'active' as indicated by Lau and Qian (2001) and Harder et al. (2002b). The sparsity of knowledge on the composition of EPS, bacterial or algal, is a limitation to understanding the role of biofilm EPS in settlement. Dissecting the various parts of the EPS puzzle will be quite a challenge since EPS represents a number of complex, heterogeneous molecules that are modified in response to environmental cues.

Diatoms have been implicated as a source of settlement cues for barnacles (Le Tourneux and Bourget 1988), abalone (Bryan and Qian 1998; Daume et al. 1999), sea urchins (Tani and Oto 1979) and sea cucumbers (Ito and Kitamura 1997), but in no case has the nature of the cues(s) been elucidated. Recent work by Qian's group has shown that settlement of larvae of *Hydroides elegans* may be induced by a diatom as well as a bacterial biofilm (Harder et al. 2002a). As with bacteria, there appeared to be a range of efficacy from inductive to non-inductive and there was no correlation with genus. EPS was implicated as the basis of the response since dead cells of inductive species also induced larval settlement (Lam et al. 2003). Extracted EPS immobilised in hydrogels from inductive species (*Achnanthes* sp. and *Nitzchia constricta*) also induced settlement (Lam et al. 2005a), glucose and galactose being the predominant sugars, the degree of induction being related to the composition of the EPS (Lam et al. 2005b). More consideration should be given to the role of algae, especially diatoms, on settlement of invertebrate larvae. Diatoms are typically 1–2 orders of magnitude larger than bacteria, many are 'motile', rapidly conditioning the surface with EPS trails, hence the impact of diatoms on settlement has probably been understated. Recent data indicate that settlement of *Balanus amphitrite* cyprids is inhibited by algal biofilms representing a number of genera (Hellio et al., unpubl.). As with bacteria, there will probably be no common effect of algae on larval settlement.

Advances in identification of microorganisms by molecular techniques will allow active strains to be studied in more detail. Work so far indicates that there are no simple patterns. For example, induction of settlement of *Hydroides elegans* larvae was triggered by more than half of the 38 strains isolated but these covered three phylogenetic branches viz *Gammaproteobacteria*, Gram positive and *Bacteriodetes* (previously the *Cytophaga-Flavobacteria-Bacteriodes* (CFB) group) (Lau et al. 2002). Furthermore, isolates from the same genus had different activities on the induction of larval settlement. A similar situation was found for *Ulva* spore settlement (see Sect. 3).

4.2
Bacterial Products and Secondary Metabolites

Secondary metabolites have been defined as naturally produced substances that are not essential to the functioning of the organisms producing them (Stone and Williams 1992). Many higher marine organisms are protected from epibiosis by the release of secondary metabolites (Clare 1996; Hay 1996; Armstrong et al. 2000; McClintock and Baker 2001; Fusetani 2004) and the composition of biofilms growing on such animate surfaces is likely to reflect the type of molecules released. Marine bacteria, cyanobacteria (blue-green algae) and biofilm algae are also rich sources of secondary products (Davidson 1995; Burja et al. 2001) and a number of studies to screen single strains of bacteria have been undertaken (Holmström and Kjelleberg 1999b) although not many of the bioactive molecules have been identified. The first fully characterised inhibitory molecule of bacterial origin was ubiquinone-8 from the culture supernatant of *Alteromonas* sp.; a strain isolated from the sponge *Halichondria okadai* (Kon-ya et al. 1995).

The major focus to identify such products stems from their potential use in biomedical applications and to lesser extent industrial applications including antifouling. Many of the products identified in such drug discovery programmes (Sennett 2001 and references therein) have not been evaluated for their role on settlement and as stated earlier, the vast majority of such studies have been conducted on cells grown in laboratory planktonic culture. The products produced by organisms growing in a biofilm are largely unexplored but recent studies (Yan et al. 2002, 2003) suggest that the production of bioactive compounds is increased when bacteria are grown as biofilms compared to shake-flask cultures.

One of the most researched strains that inhibits settlement of larvae and algal spores is *Pseudoalteromonas tunicata* strain D2 (Holmström and Kjelleberg 1999b), which belongs to a genus that is characterised by a large number of species that produce active metabolites (see Sect. 4.3).

4.3 *Pseudoalteromonas*

Pseudoalteromonas was established as a new genus by Gauthier et al. (1995), leaving the genus previously designated as *Alteromonas* with only one species. *Pseudoalteromonas* species are of interest for two reasons firstly they are commonly found in association with eukaryotic hosts in the marine environment and secondly, many species produce bioactive metabolites with antifouling properties (Holmström and Kjelleberg 1999b, 2000; Egan et al. 2000).

Many *Pseudoalteromonas* species produce a range of antibiotics including high molecular mass antibacterial compounds (Gauthier and Flatau 1976; Gauthier 1979; Gauthier and Breitmayer 1979; Baumann et al. 1984). *Pseudalteromonas luteoviolacea* produces anti-bacterial polyanionic polysaccharides (Gauthier and Flatau 1976) associated with proteins (McCarthy et al. 1994; Jiang et al. 2000) and low molecular mass brominated compounds (Gauthier and Flatau 1976; Jiang et al. 2000).

Some *Pseudoalteromonas* species produce enzymes and toxins (Holmström and Kjelleberg 1999b; Holmström et al. 2002). *P. tetradontis* produces tetrodotoxin, the neurotoxin produced by pufferfish (Simidu et al. 1990), whilst *P. piscicidia* produces a toxin that kills fish (Bein 1954). The best-known example of a *Pseudoalteromonas* species producing bioactive compounds is *P. tunicata* (Holmström et al. 1998), previously designated stain D2 and isolated from a tunicate in Swedish waters (Holmström et al. 1992). Similar dark pigmented strains have been isolated from *Ulva* in Australia (Egan et al. 2002). The production of bioactive compounds appears to be linked to the production of a yellow pigment (Egan et al. 2002) but this is not universal for all strains (Holmström et al. 2002). *P. tunicata* produces at least five extracellular compounds that act in an antifouling context by inhibiting the settlement of invertebrate larvae, algal spores and diatoms and the growth of fungi and bacteria (Holmström et al. 1996b; Holmström and Kjelleberg 1999b). The anti-bacterial compound is a large 190-kDa protein (James et al. 1996), the anti-larval molecule is small (500 Da), heat-stable and polar (Holmström and Kjelleberg 1999b) whilst the molecule that inhibits *Ulva* spores is a 3–10 kDa peptide (Holmström et al. 1996b; Egan et al. 2002). The other bioactive molecules have not yet been fully characterised. Other *Pseudoalteromonas* species have been shown to have algicidal effects (Lovejoy et al. 1998; Patel et al. 2003).

Whether the same bioactive compounds are produced when the bacterial cells are immobilised in a biofilm and what role they play in the natural environment is a subject of speculation. In this context, Ivanova et al. (1998) showed that cells cultured on hydrophilic surfaces produced more anti-microbial metabolites compared to cells cultured on hydrophobic surfaces.

5
Conclusions and Future Directions

The need for environmentally benign antifouling technologies has led to renewed interest in the ways that organisms protect themselves against predation and facilitate successful colonisation on substrata in the hostile, turbulent marine environment. Since biofilms are ubiquitous on all submerged surfaces, understanding the basis of the interactions between settling cells, spores and larvae is vital if methods are to be found that interfere with the colonisation process. However, it must be remembered that fouling organisms are opportunists and controlling one organism or even one taxon may open up niches for successful colonisation by less well known and researched species. The biofilms may themselves be the source of novel bioactive molecules. For example, enzymes derived from marine microbes may be useful in creating novel antifouling coatings by hydrolysing the sticky polymers that facilitate adhesion of spores and larvae to surfaces (Pettitt et al. 2004). Whilst natural product toxins are unlikely to be used as free compounds in coatings, they may find application as tethered (non-leaching) biocides. However, even a tethered biocide or settlement deterrent will need regulatory approval under the new European Biocidal Products Directive (98/8/CE). The costs of gaining regulatory approvals will undoubtedly inhibit progress in this area (Rittschof 2001) unless the same compounds can be exploited for other end uses (Clare 1996).

Other antifouling strategies may employ molecules that interfere with the signalling processes that influence settlement on a biofilmed surface. Two possibilities may be envisioned (Fig. 3): on the one hand, molecules that deter settlement, or their synthetic homologues, may be directly used to inhibit settlement of spores or larvae. On the other hand, molecules could be incorporated into coatings that are antagonistic to the positive signals that aid recruitment to a surface, such as antagonists of AHLs in the case of green algal spores. However, such strategies would only be

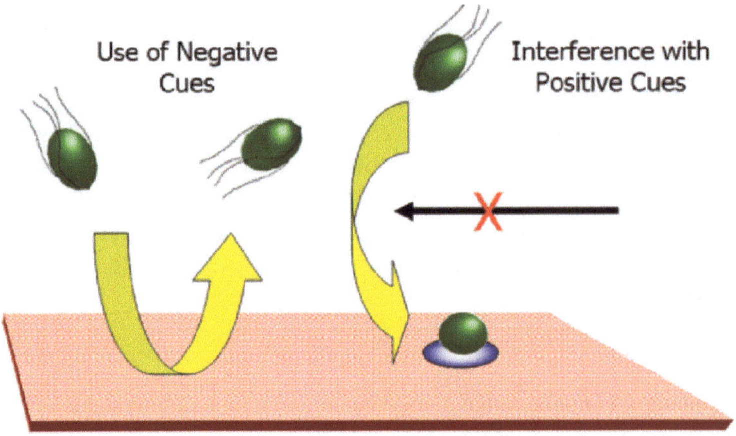

Fig. 3. Alternative approaches to control of settlement of fouling organisms (exemplified in this case by zoospores of the green alga *Ulva*), based on interference with natural signalling processes between the biofilmed surface and the spores

worth pursuing if the signals were common for a number of key fouling organisms, a scenario that may be unlikely in view of the conflicting data on biofilms amassed to date. What is certain is that regulations are continually becoming more restrictive and novel methods of fouling control need to be found. It is also unlikely that an environmentally benign 'magic bullet' will be discovered. Antifouling strategies will become more complex and the development of 'multifunctional' coatings, incorporating an array of chemistries and physical properties designed to inhibit attachment and adhesion of fouling organisms, is already being considered.

The issues that underlie the problems in unequivocally identifying the nature of biofilm-derived signals have been alluded to many times. If we are to understand more about the signals that settling cells, spores and larvae encounter and respond to, more cognisance needs to be taken of the biofilm mode of growth – a growth form that is totally different to that of bacteria in planktonic culture.

Acknowledgements. Financial support from the Natural Environmental Research Council (NER/T/S/2000/00623-5) and the United States Office of Naval Research Awards N00014-96-0373 and N00014-02-1-0311 is acknowledged.

References

Allison DG, Ruiz B, San Jose C, Jaspe A, Gilbert P (1998) Extracellular products as mediators of the formation and detachment of *Pseudomonas fluorescens* biofilms. FEMS Microbiol Lett 167:179–184

Allison DG, Gilbert P, Lappin-Scott HM, Wilson M (eds) (2000) Community structure and co-operation in biofilms. Cambridge Univ Press, Cambridge

Amann RI, Ludwig W, Schleifer KH (1995) Phylogenetic identification and in-situ detection of individual microbial cells without cultivation. Microbiol Rev 59:143–169

Amsler CD, Iken KB (2001) Chemokinesis and chemotaxis in marine bacteria and algae. In: McClintock JB, Baker BJ (eds) Marine chemical ecology. CRC Press, Boca Raton, pp 413–120

Armstrong E, Boyd KG, Burgess JG (2000) Prevention of marine biofouling using natural compounds from marine organisms. Biotechnol Annu Rev 6:221–241

Baier RE (1980) Substrata influences on the adhesion of microorganisms and their resultant new surface properties. In: Bitton G, Marshall K (eds) Adsorption of microorganisms to surfaces. Wiley, New York, pp 59–104

Bassler BL, Wright M, Silverman MR (1994) Multiple signaling systems controlling expression of luminescence in *Vibrio harveyi* – sequence and function of genes encoding a second sensory pathway. Mol Microbiol 13:273–286

Baumann P, Gauthier MJ, Baumann L (1984) Bergey's manual of determinative bacteriology. Williams and Wilkins, Baltimore

Bein SJ (1954) A study of certain chromogenic bacteria isolated from 'red tide' water with a description of a new species. Bull Mar Sci Gulf Sci 4:110–119

Biggs BJF (1996) Patterns of benthic algae of streams. In: Stevenson RJ, Bothwell M, Lowe R, Thorpe J (eds) Algal ecology: freshwater benthic ecosystem. Academic Press, London, pp 30–55

Bryan PJ, Qian P-Y (1998) Induction of larval attachment and metamorphosis in the abalone *Haliotis diversicolor* (Reeve). J Exp Mar Biol Ecol 223:39–51

Burja AM, Banaigs B, Abou-Mansour E, Burgess JG, Wright PC (2001) Marine cyanobacteria – a prolific source of natural products. Tetrahedron 57:9347–9377

Caldwell DE, Korber DR, Lawrence JR (1992) Confocal laser microscopy and digital image analysis in microbial ecology. Adv Microb Ecol 12:1–67

Callow JA, Stanley MS, Wetherbee R, Callow ME (2000) Cellular and molecular approaches to understanding primary adhesion in *Enteromorpha*: an overview. Biofouling 16:141–150

Callow JA, Osborne MP, Callow ME, Baker F, Donald AM (2002) Use of environmental scanning electron microscopy to image the spore adhesive of the marine alga *Enteromorpha* in its natural hydrated state. Colloid Surface B 27:315–521

Callow ME (2000) Algal biofilms. In: Evans LV (ed) Biofilms: recent advances in their study and control. Harwood Academic Publ, Amsterdam, pp 189–209

Callow ME, Callow JA (2000) Substratum location and zoospore behaviour in the fouling alga *Enteromorpha*. Biofouling 15:49–56

Callow ME, Callow JA, Pickett-Heaps JD, Wetherbee R (1997) Primary adhesion of *Enteromorpha* (Chlorophyta, Ulvales) propagules: quantitative settlement studies and video microscopy. J Phycol 33:938–947

Callow ME, Callow JA, Ista LK, Coleman SE, Nolasco AC, Lopez GP (2000) The use of self-assembled monolayers of different wettabilities to study surface selection and primary adhesion processes of green algal (*Enteromorpha*) zoospores. Appl Environ Microbiol 66:3249–3254

Callow ME, Jennings AR, Brennan AB, Seegert CE, Gibson A, Wilson L, Feinberg A, Baney R, Callow JA (2002) Microtopographic cues for settlement of zoospores of the green fouling alga *Enteromorpha*. Biofouling 18:237–245

Chiovitti A, Bacic A, Burke J, Wetherbee R (2003a) Heterogeneous xylose-rich glycans are associated with extracellular glycoproteins from the biofouling diatom *Craspedostauros australis* (Bacillariophyceae). Eur J Phycol 38:351–360

Chiovitti A, Higgins MJ, Harper RE, Wetherbee R, Bacic A (2003b) The complex polysaccharides of the raphid diatom *Pinnularia viridis* (Bacillariophyceae). J Phycol 39:543–554

Clare AS (1996) Marine natural product antifoulants: status and potential. Biofouling 9:211–229

Clare AS, Rittschof D, Gerhart DJ, Maki JS (1992) Molecular approaches to non-toxic antifouling. Invert Reprod Dev 22:67–76

Colwell RR, Brayton PR, Grimes DJ, Roszak DB, Huq SA, Palmer LM (1985) Viable but non-culturable *Vibrio cholerae* and related pathogens in the environment: implications for release of genetically engineered microorganisms. Bio-Technology 3:817–820

Cooksey KE, Wigglesworth-Cooksey B (1995) Adhesion of bacteria and diatoms to surfaces in the sea – a review. Aquat Microb Ecol 9:87–96

Costerton JW, Lewandowski Z, Caldwell DE, Korber DR, Lappin-Scott HM (1995) Microbial biofilms. Annu Rev Microbiol 49:711–745

Cottrell MT, Kirchman DL (2000) Community composition of marine bacterioplankton determined by 16S rRNA gene clone libraries and fluorescence in situ hybridization. Appl Environ Microbiol 66:5116–5122

Dalton HM, Poulsen LK, Halasz P, Angles ML, Goodman AE, Marshall KC (1994) Substratum-induced morphological changes in a marine bacterium and their relevance to biofilm structure. J Bacteriol 176:6900–6906

Daniel GF, Chamberlain AHL, Jones EBG (1987) Cytochemical and electron microscopical observations on the adhesive materials of marine fouling diatoms. Br Phycol J 22: 101-118

Daume S, Brand-Gardner S, Woelkerling WJ (1999) Preferential settlement of abalone larvae: diatom films vs non-geniculate coralline red algae. Aquaculture 174:243–254

Davidson BS (1995) New dimensions in natural products research: cultured marine microorganisms. Curr Opin Biotechnol 6:284–291

Davies DG, Geesey GG (1995) Regulation of the alginate biosynthesis gene algc in *Pseudomonas aeruginosa* during biofilm development in continuous-culture. Appl Environ Microbiol 61:860–867

Davies DG, Parsek MR, Pearson JP, Iglewski BH, Costerton JW, Greenberg EP (1998) The involvement of cell-to-cell signals in the development of a bacterial biofilm. Science 280:295–298

Decho AW, Moriarty DJW (1990) Bacterial exopolymer utilization by a harpacticoid copepod: a methodology and results. Limnol Oceanogr 35:1039–1049

Delong EF, Wickham GS, Pace NR (1989) Phylogenetic stains: ribosomal RNA-based probes for identification of single cells. Science 243:1360–1362

Delong EF, Franks DG, Alldredge AL (1993) Phylogenetic diversity of aggregate-attached vs free-living marine bacterial assemblages. Limnol Oceanogr 38:924–934

Dillon PS, Maki JS, Mitchell R (1989) Adhesion of *Enteromorpha* swarmers to microbial films. Microb Ecol 17:39–47

Doyle RJ (ed) (2001) Microbial growth in biofilms, vol 336. Academic Press, London

Edgar LA, Pickett-Heaps JD (1984) Diatom locomotion. Prog Phycol Res 3:47–88

Egan S, Thomas T, Holmström C, Kjelleberg S (2000) Phylogenetic relationship and antifouling activity of bacterial epiphytes from the marine alga *Ulva lactuca*. Environ Microbiol 2:343–347

Egan S, James S, Holmström C, Kjelleberg S (2002) Correlation between pigmentation and antifouling compounds produced by *Pseudoalteromonas tunicata*. Environ Microbiol 4:433–442

Eilers H, Pernthaler J, Glöckner FO, Amann R (2000) Culturability and in situ abundance of pelagic bacteria from the North Sea. Appl Environ Microbiol 66:3044–3051

Evans LV (ed) (2000) Biofilms: recent advances in their study and control. Harwood Academic Publ, Amsterdam
Ferris MJ, Ward DM (1997) Seasonal distributions of dominant 16S rRNA-defined populations in a hot spring microbial mat examined by denaturing gradient gel electrophoresis. Appl Environ Microbiol 63:1375–1381
Ferris MJ, Muyzer G, Ward DM (1996) Denaturing gradient gel electrophoresis profiles of 16S rRNA-defined populations inhabiting a hot spring microbial mat community. Appl Environ Microbiol 62:340–346
Fusetani N (2004) Biofouling and antifouling. Nat Prod Rep 21:94–104
Gardiner SM, Chhabra SR, Harty C, Williams P, Pritchard DI, Bycroft BW, Bennett T (2001) Haemodynamic effects of the bacterial quorum sensing signal molecule, *N*-(3-oxododecanoyl)-L-homoserine lactone, in conscious, normal and endotoxaemic rats. Br J Pharmacol 133:1047–1054
Garrett ES, Perlegas D, Wozniak DJ (1999) Negative control of flagellum synthesis in *Pseudomonas aeruginosa* is modulated by the alternative sigma factor AlgT (AlgU). J Bacteriol 181:7401–7404
Gauthier MJ (1979) *Alteromonas rubra* sp. nov., a new marine antibiotic-producing bacterium. Int J Syst Bacteriol 26:459–466
Gauthier MJ, Breitmayer VA (1979) A new antibiotic-producing bacterium from seawater: *Alteromonas aurantia* sp. nov. Int J Syst Bacteriol 29:366–372
Gauthier MJ, Flatau GN (1976) Antibacterial activity of marine violet-pigmented *Alteromonas* with special reference to the production of brominated compounds. Can J Microbiol 22:1612–1619
Gauthier H, Gauthier M, Christen R (1995) Phylogenetic analysis of the genera *Alteromonas*, *Shewanella* and *Moritella* using genes coding for small subunit rRNA sequences and division of the genus *Alteromonas* into genera, *Alteromonas* (emended) and *Pseudoalteromonas* gen. nov., and proposal of twelve new species combinations. Int J Syst Bacteriol 45:755–761
Geesey GG (2001) Bacterial behavior at surfaces. Curr Opin Microbiol 4:296–300
Giovannoni SJ, Britschgi TB, Moyer CL, Field KG (1990) Genetic diversity in Sargasso Sea bacterioplankton. Nature 345:60–63
Givskov M, de Nys R, Manefield M, Gram L, Maximilien R, Eberl L, Molin S, Steinberg PD, Kjellberg S (1996) Eukaryotic interference with homoserine lactone-mediated prokaryotic signalling. J Bacteriol 178:6618–6622
Glöckner FO, Fuchs BM, Amann R (1999) Bacterioplankton compositions of lakes and oceans: a first comparison based on fluorescence in situ hybridization. Appl Environ Microbiol 65:3721–3726
Granhag LM, Finlay JA, Jonsson PR, Callow JA, Callow ME (2004) Roughness-dependent removal of settled spores of the green alga *Ulva* (syn. *Enteromorpha*) exposed to hydrodynamic forces from a water jet. Biofouling 20:117–122
Hadfield MG, Paul VJ (2001) Natural chemical cues for settlement and metamorphosis of marine invertebrate larvae. In: McClintock JB, Baker BJ (eds) Marine chemical ecology. CRC Press, Boca Raton, pp 431–461
Hagström A, Pinhassi J, Zweifel UL (2000) Biogeographical diversity among marine bacterioplankton. Aquat Microb Ecol 21:231–244
Harder T, Lam C, Qian P-Y (2002a) Induction of larval settlement in the polychaete *Hydroides elegans* by marine biofilms: an investigation of monospecific diatom films as settlement cues. Mar Ecol Prog Ser 229:105–112
Harder T, Lau SCK, Dahms HU, Qian P-Y (2002b) Isolation of bacterial metabolites as natural inducers for larval settlement in the marine polychaete *Hydroides elegans* (Haswell). J Chem Ecol 28:2029–2043
Hawkins SJ, Hartnoll RG (1983) Grazing of intertidal algae by marine invertebrates. Oceanogr Mar Biol Annu Rev 21:195–283

Hay ME (1996) Marine chemical ecology: what's known and what's next? J Exp Mar Biol Ecol 200:103–134

Hayden HS, Blomster J, Maggs CA, Silva PC, Stanhope MJ, Walland RJ (2003) Linnaeus was right all along: *Ulva* and *Enteromorpha* are not distinct genera. J Phycol 38:277–294

Hentzer M, Teitzel GM, Balzer GJ, Heydorn A, Molin S, Givskov M, Parsek MR (2001) Alginate overproduction affects *Pseudomonas aeruginosa* biofilm structure and function. J Bacteriol 183:5395–5401

Hills JM, Thomason JC (1998) The effect of scales of surface roughness on the settlement of barnacle (*Semibalanus balanoides*) cyprids. Biofouling 12:57–69

Hoagland KD, Rosowski JR, Gretz MR, Roemer SC (1993) Diatom extracellular polymeric substances: function, fine structure, chemistry and physiology. J Phycol 29:537–566

Hoipkemeier-Wilson L, Schumacher JF, Carman ML, Gibson AL, Feinberg AW, Callow ME, Finlay JA, Callow JA, Brennan AB (2004) Antifouling potential of lubricious, micro-engineered, PDMS elastomers against zoospores of the green fouling alga *Ulva* (*Enteromorpha*). Biofouling 20:53–63

Holmström C, Kjelleberg S (1999a) Factors influencing the settlement of macrofoulers. In: Fingerman M, Nagabhushanam R, Thompson M-F (eds) Recent advances in marine biotechnology, vol 3. Biofilms, bioadhesion, corrosion and biofouling. Science Publ, Enfield, pp 173–202

Holmström C, Kjelleberg S (1999b) Marine *Pseudoalteromonas* species are associated with higher organisms and produce biologically active extracellular agents. FEMS Microbiol Ecol 30:285–293

Holmström C, Kjelleberg S (2000) Bacterial interactions with marine fouling organisms. In: Evans LV (ed) Biofilms: recent advances in their study and control. Harwood Academic Publ, Amsterdam, pp 101–115

Holmström C, Rittschof D, Kjelleberg S (1992) Inhibition of settlement by larvae of *Balanus amphitrite* and *Ciona intestinalis* by a surface-colonizing marine bacterium. Appl Environ Microbiol 58:2111–2115

Holmström C, James S, Egan S, Kjelleberg S (1996a) Inhibition of common fouling organisms by marine bacterial isolates with special reference to the role of pigmented bacteria. Biofouling 10:251–259

Holmström C, James S, Egan S, Kjelleberg S (1996b) Inhibition of common fouling organisms by marine bacterial isolates with special reference to the role of pigmented bacteria. Biofouling 10:251–259

Holmström C, James S, Neilan S, White D, Kjelleberg S (1998) *Pseudoalteromonas tunicata* sp. nov., a bacterium that produces antifouling agents. Int J Syst Bacteriol 48:1205–1212

Holmström C, Egan S, Franks A, McCloy S, Kjelleberg S (2002) Antifouling activities expressed by marine surface associated *Pseudoalteromonas* species. FEMS Microbiol Ecol 41:47–58

Huber B, Riedel K, Hentzer M, Heydorn A, Gotschlich A, Givskov M, Molin S, Eberl L (2001) The cep quorum-sensing system of *Burkholderia cepacia* H111 controls biofilm formation and swarming motility. Microbiol Res 147:2517–2528

Hudon C, Bourget E, Legendre P (1983) An integrated study of the factors influencing the choice of the settling site of *Balanus crenatus* cyprid larvae. Can J Fish Aquat Sci 40:1186–1194

Imhoff JF, Stohr R (2003) Sponge-associated bacteria: general overview and special aspects of bacteria associated with *Halichondria panicea*. In: Müller WEG (ed) Marine molecular biotechnology. Sponges (Porifera). Springer, Berlin Heidelberg New York, pp 35–57

Ista LK, Callow ME, Finlay JA, Coleman SE, Nolasco AC, Simons RH, Callow JA, Lopez GP (2004) Effect of substratum surface chemistry and surface energy on attachment of marine bacteria and algal spores. Appl Environ Microbiol 70:4151–4157

Ito Y, Kitamura H (1997) Induction of larval metamorphosis in the sea cucumber, *Stichopus japonicus* by periphytic diatoms. Hydrobiologia 358:281–284

Ivanova E, Nicolau DV, Yumoto N, Taguchi T, Okamoto K, Tatsu Y, Yoshikawa S (1998) Impact of conditions of cultivation and adsorption on antimicrobial activity of marine bacteria. Mar Biol 130:545–551

Jackson SM, Jones EBG (1988) Fouling film development on antifouling paints with special reference to film thickness. Int Biodeteri 24:277–287

James S, Holmström C, Kjelleberg S (1996) Purification and characterisation of a novel antibacterial protein from the marine bacterium D2. Appl Environ Microbiol 62:2783–2788

Jiang Z, Boyd KG, Mearns-Spragg A, Adams DR, Wright PC, Burgess JG (2000) Two diketopiparazines and one halogenated phenol from cultures of the marine bacterium *Pseudoalteromonas luteoviolacea.* Nat Prod Lett 14:435–440

Joint I, Callow ME, Callow JA, Clarke KR (2000) The attachment of *Enteromorpha* zoospores to a bacterial biofilm assemblage. Biofouling 16:151–158

Joint I, Tait K, Callow ME, Callow JA, Milton D, Williams P, Camara M (2002) Cell-to-cell communication across the prokaryote-eukaryote boundary. Science 298:1207

Keough MJ, Raimondi PT (1995) Responses of settling invertebrate larvae to bioorganic films: effects of different types of films. J Exp Mar Biol Ecol 235–253

Keough MJ, Raimondi PT (1996) Responses of settling invertebrate larvae to bioorganic films: effects of large-scale variation in films. J Exp Mar Biol Ecol 207:59–68

Khandeparker L, Anil AC, Raghukumar S (2003) Barnacle larval destination: piloting possibilities by bacteria and lectin interaction. J Exp Mar Biol Ecol 289:1–13

Kirchman D, Graham S, Reish D, Mitchell R (1982a) Lectins may mediate settlement and metamorphosis of *Janua (Dexiospira) brasiliensis* Grube (Polychaeta: Spirorbidae). Mar Biol Lett 3:131–138

Kirchman D, Graham S, Reish D, Mitchell R (1982b) Bacteria induce settlement and metamorphosis of *Janua* (*Dexiospira*) Grube (Polychaeta:Spirorbideae). J Exp Mar Biol Ecol 56:153–163

Kon-ya K, Schimidzu N, Otaki N, Yokoyama A, Adachi K, Miki W (1995) Inhibitory effects of bacterial ubiquinones on the settling of the barnacle, *Balanus amphitrite.* Experientia 51:153–155

Lam C, Harder T, Qian P-Y (2003) Induction of larval settlement in the polychaete *Hydroides elegans* by surface-associated settlement cues of marine diatoms. Mar Ecol Prog Ser 263:83–92

Lam C, Harder T, Qian P-Y (2005a) Induction of larval settlement in the polychaete *Hydroides elegans* by extracellular polymers of benthic diatoms. Mar Ecol Prog Ser 286:145–154

Lam C, Harder T, Qian P-Y (2005b) Different growth conditions of pennate diatoms affect the quality and quantity of extracellular polymers and their role as larval settlement cues. Mar Ecol Prog Ser (in press)

Lappin-Scott HM, Costerton JW (1995) Microbial biofilms. Cambridge Univ Press, Cambridge

Lau SCK, Qian P-Y (2000) Inhibitory effect of phenolic compounds and marine bacteria on larval settlement of the barnacle *Balanus amphitrite* Darwin. Biofouling 16:47–58

Lau SCK, Qian P-Y (2001) Larval settlement in the serpulid polychaete *Hydroides elegans* in response to bacterial films: an investigation of the nature of putative larval settlement cue. Mar Biol 138:321–328

Lau SCK, Mak KKW, Chen F, Qian P-Y (2002) Bioactivity of bacterial strains isolated from marine biofilms in Hong Kong waters for the induction of larval settlement in the marine polychaete *Hydroides elegans.* Mar Ecol Prog Ser 226:301–310

Lau SC, Harder T, Qian P-Y (2003a) Induction of larval settlement in the serpulid polychaete *Hydroides elegans* (Haswell): role of bacterial extracellular polymers. Biofouling 19:197–204

Lau SCK, Thiyagarajan V, Qian P-Y (2003b) The bioactivity of bacterial isolates in Hong Kong waters for the inhibition of barnacle (*Balanus amphitrite* Darwin) settlement. J Exp Mar Biol Ecol 282:43–60
Lawrence RN, Dunn WR, Bycroft BW, Cámara M, Chhabra SR, Williams P, Wilson VG (1999) The *Pseudomonas aeruginosa* quorum sensing signal molecule, N-(3-oxododecanoyl)-L-homoserine lactone inhibits porcine arterial smooth muscle contraction. Br J Pharmacol 128:845–848
Le Tourneux F, Bourget E (1988) Importance of physical and biological settlement cues used at different spatial scales by the larvae of *Semibalanus balanoides*. Mar Biol 97:57–66
Lewandowski Z (2000) Structure and function of biofilms. In: Evans LV (ed) Biofilms: recent advances in their study and control. Harwood Academic Publ, Amsterdam, pp 1–18
Lind JL, Heimann K, Miller EA, van Vliet C, Hoogrenradd NJ, Wetherbee R (1997) Substratum adhesion and gliding in a diatom are mediated by estracellular proteoglycans. Planta 203:213–221
Little B, Zsolnay A (1985) Chemical fingerprinting of adsorbed organic materials on metal surfaces. J Colloid Interface Sci 104:79–86
Lock MA (1981) River epilithon – a light and organic energy transducer. In: Lock MA, Williams DD (eds) Perspectives in running water ecology. Plenum Press, New York, pp 3–39
Loeb GI, Neihof RA (1975) Marine conditioning films. Adv Chem 145:319–335
Lovejoy C, Bowman JP, Hallegraeff GM (1998) Algicidal effects of a novel marine *Pseudoalteromonas* isolate (class Proteobacteria, gamma subdivision) on harmful algal bloom species of the genera *Chattonella, Gymnodinium*, and *Heterosigma*. Appl Environ Microbiol 64:2806–2813
Maki JS (1999) The influence of marine microbes on biofouling. In: Fingerman M, Nagabhushanam R, Thompson M-F (eds) Recent advances in marine biotechnology, vol 3. Biofilms, bioadhesion, corrosion and biofouling. Science Publ, Enfield, pp 141–171
Maki JS, Rittschof D, Costlow JD, Mitchell R (1988) Inhibition of attachment of larval barnacles, *Balanus amphitrite*, by bacterial surface films. Mar Biol 97:199–206
Maki JS, Rittschof D, Samuelsson M-O, Szewzyk U, Yule AB, Kjelleberg S, Costlow JD (1990) Effect of marine bacteria and their exopolymers on the attachment of barnacle cypris larvae. Bull Mar Sci 46:499–511
Manefield M, de Nys R, Kumar N, Read R, Givskov M, Steinberg P, Kjelleberg SA (1999) Evidence that halogenated furanones from *Delisea pulchra* inhibit acylated homoserine lactone (AHL)-mediated gene expression by displacing the AHL signal from its receptor protein. Microbiology 145:283–291
Marshall KC (1996) Adhesion as a strategy for access to nutrients. In: Fletcher M (ed) Bacterial adhesion: molecular and ecological diversity. Wiley, New York, pp 59–87
Mathesius U, Mulders S, Gao M, Teplitski M, Caetano-Anolles G, Rolfe BG, Bauer WD (2003) Extensive and specific responses of a eukaryote to bacterial quorum-sensing signals. Proc Natl Acad Sci USA 100:1444–1449
Matsuo Y, Suzuki M, Kasai H, Shizuri Y, Harayama S (2003) Isolation and phylogenetic characterization of bacteria capable of inducing differentiation in the green alga *Monostroma oxyspermum*. Environ Microbiol 5:25–35
McCarthy S A, Johnson R M, Kakimoto D (1994) Characterization of an antibiotic produced by *Alteromonas luteoviolacea* Gauthier 1982, 85 isolated from Kinko Bay, Japan. J Appl Bacteriol 77:426–432
McClintock JB, Baker BJ (eds) (2001) Marine chemical ecology. CRC Press, Boca Raton
Miller MB, Bassler BL (2001) Quorum sensing in bacteria. Annu Rev Microbiol 55:165-199
Milton DL, Hardman A, Camara M, Chhabra SR, Bycroft BW, Stewart G, Williams P (1997) Quorum sensing in *Vibrio anguillarum*: Characterization of the vanI/vanR locus and

identification of the autoinducer *N*-(3-oxodecanoyl)-L-homoserine lactone. J Bacteriol 179:3004–3012
Milton DL, Chalker VJ, Kirke D, Hardman A, Camara M, Williams P (2001) The luxM Homologue vanM from *Vibrio anguillarum* directs the synthesis of *N*-(3-hydroxyhexanoyl)homoserine lactone and *N*-hexanoylhomoserine lactone. J Bacteriol 183:3537-3547
Mitchell JG, Pearson L, Dillon S (1996) Cluster dynamics of marine bacteria in seawater enrichments. Appl Environ Microbiol 62:3716–3721
Muyzer G (1999) DGGE/TGGE a method for identifying genes from natural ecosystems. Curr Opin Microbiol 2:317–322
Nakanishi K, Nishijima M, Nishimura M, Kuwano K, Saga N (1996) Bacteria that induce morphogenesis in *Ulva pertusa* (Chlorophyta) grown under axenic conditions. J Phycol 32:479–482
Nakanishi K, Nishijima M, Nomoto AM, Yamazaki A, Saga N (1999) Requisite morphologic interaction for attachment between *Ulva pertusa* (Chlorophyta) and symbiotic bacteria. Mar Biotechnol 1:107–111
Nivens DE, Ohman DE, Williams J, Franklin M (2001) Role of alginate and its O-acetylation in formation of *Pseudomonas aeruginosa* microcolonies and biofilms. J Bacteriol 183:1047–1057
Oliver JD, Nilsson L, Kjelleberg S (1991) Formation of nonculturable *Vibrio vulnificus* cells and its relationship to the starvation state. Appl Environ Microbiol 57:2640–2644
Olivier F, Tremblay R, Bourget E, Rittshof D (2000) Barnacle settlement: field experiments on the influence of larval supply, tidal level, biofilm quality and age on *Balanus amphitrite* cyprids. Mar Ecol Prog Ser 199:185–204
Patel P, Callow ME, Joint I, Callow JA (2003) Specificity in the settlement - modifying response of bacterial biofilms towards zoospores of the marine alga *Enteromorpha.* Environ Microbiol 5:338–349
Pawlik JR (1992) Chemical ecology of the settlement of benthic marine invertebrates. Oceanogr Mar Biol Annu Rev 30:273–335
Pettitt ME, Henry SL, Callow ME, Callow JA, Clare AS (2004) Activity of commercial enzymes on settlement and adhesion of cypris larvae of the barnacle *Balanus amphitrite*, spores of the green alga *Ulva linza*, and the diatom *Navicula perminuta.* Biofouling 20:299–311
Provasoli L, Pintner IJ (1980) Bacteria induced polymorphism in axenic laboratory strain of *Ulva lactuca* (Chlorophyceae). J Phycol 16:196–201
Railkin AI (2004) Marine biofouling: colonization processes and defences. CRC Press, Boca Raton
Rittschof D (2001) Natural product antifoulants and coatings development. In: McClintock JB (ed) Marine chemical ecology. CRC Press, Boca Raton
Roberts D, Rittschof D, Holm E, Schmidt AR (1991) Factors influencing initial larval settlement: temporal, spatial and surface molecular components. J Exp Mar Biol Ecol 150:203–211
Roszak DB, Grimes DJ, Colwell RR (1984) Viable but nonrecoverable stage of *Salmonella enteritidis* in aquatic systems. Can J Microbiol 30:334–338
Round FE, Crawford RM, Mann DG (1990) The diatoms. Cambridge Univ Press, Cambridge
Schmidt TM, Delong EF, Pace NR (1991) Analysis of a marine picoplankton community by 16S ribosomal RNA gene cloning and sequencing. J Bacteriol 173:4371–4378
Sennett SH (2001) Marine chemical ecology: applications in marine biomedical prospecting. In: McClintock JB, Baker BJ (eds) Marine chemical ecology. CRC Press, Boca Raton, pp 523–542
Simidu U, Kita-Tsukamoto K, Yasumoto T, Yotsu M (1990) Taxonomy of four marine bacterial strains that produce tetrodotoxin. Int J Syst Bacteriol 48:1205–1212

Stanley MS, Callow ME, Callow JA (1999) Monoclonal antibodies to adhesive cell coat glycoproteins secreted by zoospores of the green alga *Enteromorpha*. Planta 210:61–71
Steinberg PD, de Nys R (2002) Chemical mediation of colonization of seaweed surfaces. J Phycol 38:621–629
Steinberg PD, Schneider R, Kjelleberg S (1997) Chemical defences of seaweeds against microbial colonization. Biodegradation 8:211–220
Steinberg PD, de Nys R, Kjelleberg S (1998) Chemical inhibition of epibiota by Australian seaweeds. Biofouling 12:227–244
Steinberg PD, de Nys R, Kjellberg S (2001) Chemical mediation of surface colonization. In: McClintock JB, Baker BJ (eds) Marine chemical ecology. CRC Press, Boca Raton, pp 355–387
Steinberg PD, de Nys R, Kjellberg S (2002) Chemical cues for surface colonization. J Chem Ecol 28:1935–1951
Stone MJ, Williams DH (1992) On the evolution of functional secondary metabolites (natural products). Mol Microbiol 6:29–34
Stoodley P, Lewandowski Z, Boyle JD, Lappin-Scott HM (1999) Structural deformation of bacterial biofilms caused by short- term fluctuations in fluid shear: an in situ investigation of biofilm rheology. Biotechnol Bioeng 65:83–92
Stoodley P, Sauer K, Davies DG, Costerton JW (2002) Biofilms as complex differentiated communities. Annu Rev Microbiol 56:187–209
Sutherland IW (2001a) Biofilm exopolysaccharides: a strong and sticky framework. Microbiology 147:3–9
Sutherland IW (2001b) The biofilm matrix – an immobilized but dynamic microbial environment. Trend Microbiol 9:222–227
Swift S, Downie JA, Whitehead NA, Barnard AML, Salmond GPC, Williams P (2001) Quorum sensing as a population-density-dependent determinant of bacterial physiology. Adv Microb Physiol 45:199–270
Taga ME, Bassler BL (2003) Chemical communication among bacteria. Proc Natl Acad Sci USA 100:14549–14554
Tait K, Joint I, Daykin M, Milton D, Williams P, Camara M (2005) Disruption of quorum sensing in seawater abolishes attraction of zoospores of the green alga to bacterial biofilms. Environ Microbiol 7:229–240
Tani Y, Oto Y (1979) Effects of benthic diatoms on settlement and metamorphosis of the sea urchin *Pseudocentrotus depressus*. Suisanzoshoku 27:148–150
Tatewaki M (1983) Morphogenesis of *Monostroma oxyspermum* (Kutz) Doty (Chlorophyceae) in axenic culture, especially in bialgal culture. J Phycol 19:409–416
Telford G, Wheeler D, Williams P, Tomkins PT, Appleby P, Sewell H, Stewart GSAB, Bycroft BW, Pritchard DI (1998) The *Pseudomonas aeruginosa* quorum sensing signal molecule, *N*-(3-oxododecanoyl)-L-homoserine lactone has immunomodulatory activity. Infect Immun 66:36–42
Teplitski M, Chen H, Rajamani S, Gao M, Merighi M, Sayre RT, Robinson JB, Rolfe BG, Bauer WD (2004) *Chlamydomonas reinhardii* secretes compounds that mimic bacterial signals and interfere with quorum sensing regulation in bacteria. Plant Physiol 134:137–146
Thomas RWSP, Allsop D (1983) The effects of certain periphytic marine bacteria upon the settlement and growth of *Enteromorpha*, a fouling alga. Biodeterioration 5:348–357
Thompson RC, Norton TA, Hawkins SJ (1998) The influence of epilithic microbial films on the settlement of *Semibalanus balanoides* cyprids – a comparison between laboratory and field experiments. Hydrobiologia 375/376:203–216
Todd CD, Keough MJ (1994) Larval settlement in hard substratum epifaunal assemblages – a manipulative field-study of the effects of substratum filming and the presence of incumbents. J Exp Mar Biol Ecol 181:159–187

Tsurumi K, Fusetani N (1998) Effects of early fouling communities formed in the field on settlement and metamorphosis of cyprids of the barnacle *Balanus amphitrite* Darwin. Biofouling 12:221–228

Van den Hoek C, Mann DG, Jahns HM (1995) Algae: an introduction to phycology. Cambridge Univ Press, Cambridge

Vandevivere P, Kirchman D L (1993) Attachment stimulates exopolysaccharide synthesis by a bacterium. Appl Environ Microbiol 59:3280–3286

Wahl M (1989) Marine epibiosis.1. Fouling and antifouling - some basic aspects. Mar Ecol Prog Ser 58:175–189

Wetherbee R, Lind JL, Burke J (1998) The first kiss: establishment and control of initial adhesion by raphid diatoms. J Phycol 34:9–15

Whiteley M, Bangera MG, Bumgarner RE, Parsek MR, Teitzel GM, Lory S, Greenberg EP (2001) Gene expression in *Pseudomonas aeruginosa* biofilms. Nature 413:860–864

Wieczorek SK, Todd CD (1997) Inhibition and facilitation of bryozoan and ascidian settlement by natural multi-species biofilms: effects of film age and the roles of active and passive larval attachment. Mar Biol 128:463–473

Wieczorek SK, Todd CD (1998) Inhibition and facilitation of settlement of epifaunal marine invertebrate larvae by microbial biofilm cues. Biofouling 12:81–118

Wieczorek SK, Clare AS, Todd CD (1995) Inhibitory and facilitatory effects of microbial films on settlement of *Balanus amphitrite amphitrite* larvae. Mar Ecol Prog Ser 119:221–228

Wimpenny J (2000) An overview of biofilms as functional communities. In: Allison D, Gilbert P, Lappin-Scott HM, Wilson M (eds) Community structure and co-operation in biofilms. Cambridge Univ Press, Cambridge, pp 1–24

Wimpenny J, Manz W, Szewzyk U (2000) Heterogeneity in biofilms. FEMS Microbiol Rev 24:661–671

Withers H, Swift S, Williams P (2001) Quorum sensing as an integral component of gene regulatory networks in Gram-negative bacteria. Curr Opin Microbiol 4:186–189

Wustman BA, Gretz MR, Hoagland KD (1997) Extracellular matrix assembly in diatoms (Bacillariophyceae) I. A model of adhesives based on chemical characterization and localization of polysaccharides from the marine diatom *Achnanthes longipes* and other diatoms. Plant Physiol 113:1059–1069

Yan LM, Boyd KG, Burgess JG (2002) Surface attachment induced production of antimicrobial compounds by marine epiphytic bacteria using modified roller bottle cultivation. Mar Biotechnol 4:356–366

Yan LM, Boyd KG, Adams DR, Burgess JG (2003) Biofilm-specific cross-species induction of antimicrobial compounds in bacilli. Appl Environ Microbiol 69:3719–3727

Zobell CE (1943) The effect of solid surfaces on bacterial activity. J Bacteriol 46:39–56

Zobell CE, Allen EC (1935) The significance of marine bacteria in the fouling of submerged surfaces. J Bacteriol 29:239–251

Adrenoceptor and Other Pharmacoactive Compounds as Putative Antifoulants

M. Dahlström, H. Elwing

Abstract. The search for new antifouling methods, which are non-hazardous for the marine environment, is intense. However, even if several innovations in this field of research have been made, the search for unique molecules with characteristics such as strong biological activity, low residence time in the marine environment and which target special physiological features in marine invertebrate larvae, biofilm forming bacteria or algal spores is still required. This chapter reviews the effects of biogenic amine receptor agonists and antagonists, primarily G protein-coupled receptors, on settling barnacle cypris larvae. Biotechnological research on adrenoceptor compounds as lead molecules in new antifouling technologies is also reviewed.

1 Introduction

Larvae of sessile marine invertebrates that settle and form dense communities on man-made structures, e.g., boats' and ship' hulls cause severe economical and ecological problems. Economically, by loss of speed, increased fuel consumption, increased time spent in dry-docking and ecologically, by transmitting organisms into new areas where they pose a potential threat to endemic species. Hitherto, this problem has been overcome by using highly toxic paints based on heavy metals, such as tributyltin. However, the adverse effects of the organotin compounds on the marine environment (Alzieu et al. 1986; Gibbs et al. 1990) and human health concerns can no longer be overlooked (Clare et al. 1992). Therefore, the IMO has recently passed a resolution (IMO 1999) that encompasses a complete ban on the use of tin-based antifouling paints. This has encouraged increased research efforts to understand and characterize underlying molecular events in the settlement process of both invertebrate larvae and algal spores, to guide the search for novel,

M. Dahlström
Department of Marine Ecology, Tjärnö Marine Biological Laboratory, Göteborg University, SE-452 96 Strömstad, Sweden

H. Elwing
Department of Cell and Molecular Biology, Göteborg University, Box 462, SE-405 30 Göteborg, Sweden

Progress in Molecular and Subcellular Biology
Subseries Marine Molecular Biotechnology
N. Fusetani, A.S. Clare (Eds.): Antifouling Compounds

non- or low-toxic, target-specific substances (Clare et al. 1992; Clare and Matsumura 2000).

Among the marine fouling species, the barnacle is particularly prominent due to its strong attachment to the surface, its calcareous, hard body amour and its relatively high resistance to toxic components leaking from paint films (Rittschof 2000). Many research groups have therefore chosen the barnacle as a model fouling organism. The barnacle cypris larva displays a complex exploratory behavior before it permanently attaches itself to the substratum after which metamorphosis is completed (Crisp 1961; Berntsson et al. 2000). The exploratory phase is a sensitive stage in the settlement process, which might easily be disrupted by disturbances of a chemical and mechanical nature. The aim of the present review is to describe the current status of adrenoceptor and other pharmacoactive compounds used in the research of cypris larval signaling pathways, as well as their application in antifouling research.

1.1 Some Basic Aspects of Pharmacoactive Compounds

Pharmacoactive compounds are used for therapeutic purposes, mainly in mammals, to cure or relieve a wide array of pathological conditions. They are also widely used as tools in pharmacological research to characterize receptors and signaling pathways. The main target of a pharmacoactive compound is a protein, even if some drugs exist which interact with target sites on DNA (antitumor and antimicrobial drugs). There are four main kinds of regulatory proteins to which pharmacoactive compounds bind and exert their action: (1) receptor proteins; (2) enzymes; (3) ion channels and (4) carrier proteins. Still, many of the drugs in use today, have unknown sites of action and their mechanisms remain to be elucidated. It is estimated that 60% of the drugs prescribed today act on G protein-coupled receptors (Wilson et al. 1998) and that at least 50% of the total drugs in use have a natural origin (Newman et al. 2003). Classically, the isolation and characterization of pharmacoactive compounds have thus been guided by their effect on vertebrates, in particular mammals. Their further improvement by rational drug design, e.g., through structure-activity relationships, release characteristics and targeting, is also based on their classification in vertebrate model systems. However, growing knowledge of signal transduction systems in invertebrates suggest that neuropharmacological compounds with a human medicinal history, e.g., cocaine, are secondary metabolites synthesized by plants as natural insecticides (Nathanson et al. 1993). In humans, cocaine acts by blocking dopamine re-uptake while in insects, cocaine affects octopaminergic uptake, and octopamine transmission (Nathanson et al. 1993). Octopamine is a major biogenic amine in invertebrates (Roeder 1999) but is only found in trace amounts in vertebrates (Borowsky et al. 2001). Specific octopamine receptors are putative targets for novel insecticides (reviewed by Roeder 1999, 2002). It is also currently recognized that invertebrate receptors, e.g., biogenic amine receptors, display different pharmacological profiles, even if they bind the same transmitters as their vertebrate counterparts (Hen 1992; Tierney 2001). This difference can prove useful for the use of synthetic agonists or antagonists to these receptors in the management of, for example, arthropod pests.

The appealing thought of specific endogenous targets present only in weeds, insect pests, fungi or pathogens affecting human activities, has long led to attempts to design biocides, or drugs, with little or no effects on non-target species. For example, the recent discovery of the shikimate pathway, responsible for the biosynthesis of, for example, tryptophan, in the apicomplexan parasite, *Plasmodium falciparum*, has presented a new means to combat malaria caused by this parasitic protozoan (Roberts et al. 1998; Fitzpatrick et al. 2001).

Accumulating knowledge on physiological mechanisms and signaling pathways in marine invertebrate larvae constitutes a new means to rationally design molecules, which interfere with settlement and metamorphosis in attempts to combat marine biofouling (Clare et al. 1992; Clare and Matsumura 2000). This chapter will concentrate on molecules known to interact with vertebrate/mammalian biogenic amine G protein-coupled receptors and the possibility that they could be used as marine biocides against the larvae of hard-fouling organisms. The emphasis will be on biogenic amine signaling and pharmacoactive compounds, which in experimental studies have indicated an interaction with these systems. We will also present results of the biotechnological approach of our own research concerning adrenoceptor-active compounds targeted against barnacles.

2 G Protein-Coupled Receptors

Receptors mediate intra- and intercellular communication and ultimately, assist in regulating physiological processes. Receptors can largely be divided into four superfamilies, one of which is the G protein-coupled receptors (GPCRs), also known as seven transmembrane receptors. These receptors are present in yeasts and molds as well as in mammals (Peroutka 1994). The other three superfamilies of receptors are ion-channel receptors (ionotropic receptors), kinase-linked receptors and nuclear receptors. Many neurotransmitters, neuromodulators, neurohormones, local mediators and chemical and physical (e.g., photons) stimuli exert their action by binding to GPCRs (Brody and Cravchik 2000). More than a thousand GPCRs are known and more are continuously being discovered (Hamm 1998).

The GPCR superfamily consists of membrane-bound receptors that are highly conserved in their structure and despite their diversity of endogenous ligands, share a similar, basic, protein scaffold (Hamm 1998; Meeusen et al. 2003). They all consist of seven transmembrane α-helices which carry the most conserved regions. The transmembrane regions are connected extracellularly and intracellularly via three loops. The N-terminus is situated extracellularly while the C-terminus is placed intracellularly (Fig. 1). Only quite recently was the first crystal structure of a G protein revealed (Palczewski et al. 2000).

Biogenic amines interact with a group of GPCRs called the rhodopsin-like receptors which is the largest subfamily of GPCRs. The binding domain for small molecules like the biogenic amines appears to be situated within the cleft between the α-helical segments in the plane of the membrane (Blenau and Baumann 2001). Ligand binding to GPCRs activates the receptor and causes changes in the relative orientation of transmembrane helices 3 and 6 (Hamm 1998) (Fig. 1). These changes in turn, cause binding sites on the intracellular loops, engaged in G protein binding, to be exposed (Wess 1997). The interaction of the receptor

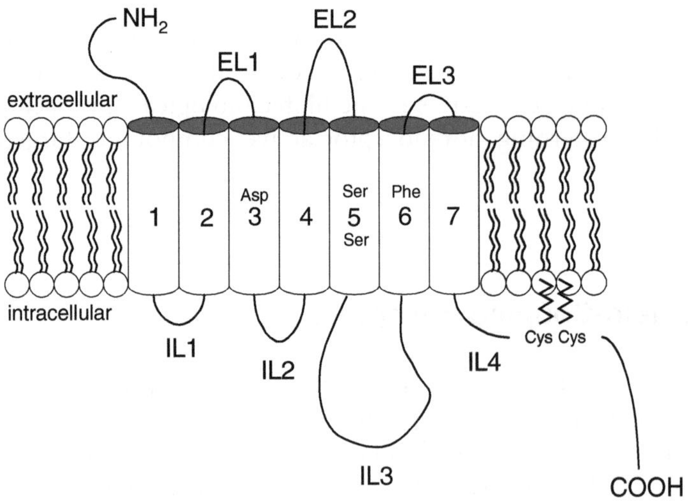

Fig. 1. A schematic illustration of a G protein-coupled receptor (GPCR). The receptor protein spans the membrane seven times as illustrated by the cylinders in the drawing. Amino acid residues that take part in ligand binding are aspartate (Asp) in TM3, serine (Ser) residues in TM5 and phenylalanine (Phe) in TM6. A fourth intracellular loop (IL4) is formed when the protein is posttranslatioinally palmitoylated at cysteine (Cys) residues in the intracellular tail. Redrawn from Blenau and Baumann (2001)

cytoplasmic loops with the heterotrimeric G protein causes the exchange of bound GDP to GTP, thus activating the G protein. Upon GTP binding, the G protein, consisting of α, β and γ-subunits, dissociates into a α-GTP and a βγ-subunit. These subunits are the active forms of the G protein and interact with various effector molecules such as enzymes and ion channels. The process of G protein signaling is concluded by the hydrolysis of GTP to GDP on the α-subunit. Moreover, receptor firing is 'turned off' by the phosphorylation of specific amino acid residues in the C-terminus and the third intracellular loop (Palczewski 1997).

The intracellular signal pathway activated by the interaction of G protein subunits to effector molecules results in a change in the intracellular concentration of cyclic AMP and/or Ca^{2+}. Four main classes of G proteins are known, namely G_s, G_i, G_q and, G_{12} and G_{13}. It is well documented that G_s stimulates adenylyl cyclase, leading to increased cellular concentrations of cyclic AMP, while G_i inhibits adenylyl cyclase. G_q activates phospholipase C, leading to the formation of the two second-messengers, inositol (1,4,5) tris-phosphate (IP_3) and diacylglycerol (DAG). The functions of G_{12} and G_{13} are so far unknown (Gudermann et al. 1996; Hamm 1998). Hence, when the receptor binds a G_s protein, this will lead to increased cellular concentrations of cyclic AMP. This in turn activates various protein kinases, involved in phosphorylation processes. Protein kinases affect cytosolic proteins and ion channel receptors, as well as transcription factors (de Cesare et al. 1999). Cyclic AMP is involved in regulating energy metabolism, e.g., increased lipolysis, reduced glycogen synthesis, cell division, and cell differentiation.

A common theme for the regulation of settlement and metamorphosis of marine invertebrate larvae has long been sought, since their metamorphosis differs fundamentally from their terrestrial counterparts (see Hadfield 2000 for an excellent review). Indeed, settling larvae of several marine species utilize external chemical signals as inducers for attachment and metamorphosis (Rittschof 1985; Hadfield and Pennington 1990; Morse 1990; Leitz 1998; Clare and Matsumura 2000). However, the nature of the receptors to which these inducers bind, to evoke the metamorphic response, is unknown. In barnacle larvae there is evidence for the involvement of a G protein-coupled receptor(s) (Rittschof et al. 1986; Clare et al. 1995), while in for example the polychaete, *Hydroides elegans*, settlement is not modulated by GPCRs (Holm et al. 1998). Accumulating sequence information through the large genome projects, of which the *Drosophila melanogaster* and the *Caenorhabditis elegans* projects were the first invertebrate genomes to become fully mapped, has made it possible to search for sequence similarities in, for example, GPCRs. Where marine fouling organisms are concerned, there is evidence that at least two GPCRs are present in the barnacle, *Balanus amphitrite*, namely, G protein-coupled adrenergic and serotonergic receptors (Isoai et al. 1996; Kawahara et al. 1997). Considering the range of physiological processes that GPCRs mediate, including chemosensation, and the recent data that suggests that up to 1.6% of the mosquito, *Anopheles gambiae*, genome consists of sequences for GPCRs (Hill et al. 2002), more GPCRs are likely to be present in barnacle larvae.

3 Biogenic Amine Signaling and Implications in the Settlement of Barnacle Larvae

The classical transmitter molecules are believed to have evolved around 1 billion years ago and are present in all animal phyla, except possibly in the Porifera and the Cnidaria (Walker et al. 1996). Transmitters belonging to the class called biogenic amines have attracted much attention since the finding that they are responsible for the generation and regulation of a range of behaviors (Bicker and Menzel 1989). The eight known biogenic amines are subdivided into indoleamine (serotonin), catecholamines (dopamine, noradrenaline and adrenaline), monoamines (octopamine and tyramine), imidazoleamine (histamine) and amino acid (γ-amino butyric acid (GABA)) (Fig. 2) and are derived from the decarboxylation of amino acids (Roeder 2002).

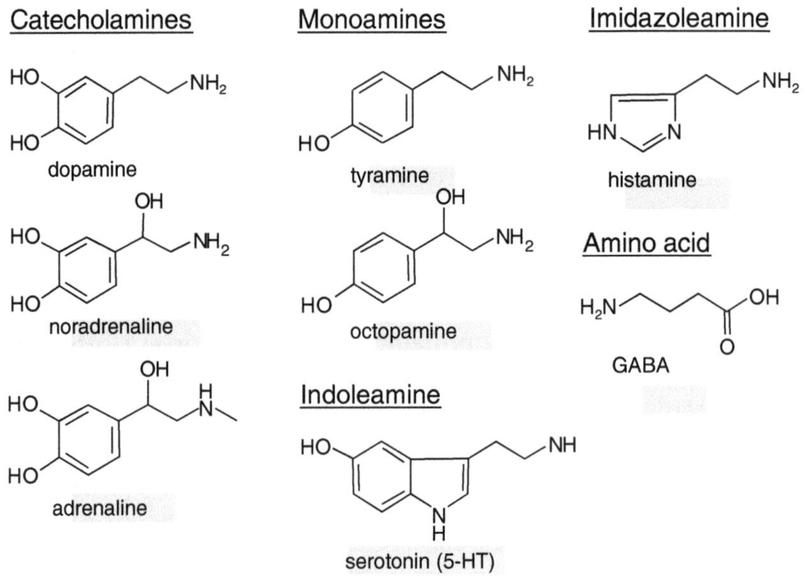

Fig. 2. The chemical structures of the biogenic amines. Redrawn from Roeder (2002)

In invertebrates, the major biogenic amines are dopamine, tyramine, serotonin, octopamine and histamine (Blenau and Baumann 2001). Biogenic amines act in any of three basic modes depending on the system: 1) as classical neurotransmitters; 2) as neuromodulators; or 3) as neurohormones (Weiger 1997; Roeder 1999; Libersat and Pflueger 2004). Neuromodulators are released into a broader area than neurotransmitters and their target is often a G-protein-coupled receptor. At the molecular level, neuromodulators affect different functions such as modulation of ion channels and receptors, protein synthesis, enzyme activity and gene transcription. Unlike neurotransmitters, which act rapidly to effect the accurate transmission of the nerve signal, without losses in signal strength, the effects of neuromodulators may last for weeks (Libersat and Plueger 2004). For example, neuromodulation by serotonin and octopamine has been shown to evoke an increase in sensitivity to environmental signals, such as pheromones that are employed in the search for mating partners (Birmingham and Tauck 2004; Gattelier et al. 2004). Neuromodulators control an array of complex behaviors in many invertebrates (insects and to a lesser extent crustaceans have received most attention), including feeding, aggression and flight (Weiger 1997; Beltz 1999; Roeder 1999; Libersat and Pflueger 2004).

In many invertebrates, biogenic amines can also act as neurohormones through their release into the circulatory system with consequences for the entire organism. For example, the catecholamines, dopamine and noradrenaline, have recently been described as neuroendocrine messengers in the stress-induced modulation of the immune system in mollusks (Lacoste et al. 2002). Furthermore, it has been found that aspects of reproduction such as egg laying in the nematode, *C. elegans*, are regulated by serotonin and dopamine (Weinshenker et al. 1995).

Below we review the evidence that implicates the different biogenic amines in barnacle larval attachment and metamorphosis. Most studies concerning barnacle larvae have been performed using whole animal experiments where the different biogenic amines, as well as agonists and antagonists to these systems, have been applied exogenously in the water of the test container. Such conditions are of course not ideal for studying internal signal transduction (see Pawlik 1990), or for understanding the mechanism of action of antifouling compounds. However, we here apply the pharmacological interpretation, as suggested by Clare and Matsumura (2000), where the settlement process of barnacle larvae can be separated pharmacologically into attachment and metamorphosis.

3.1 Serotonin (5-Hydroxytryptamine (5-HT))

Serotonergic neurons are present in all animal phyla with a nervous system (Weiger 1997) and it has been suggested that serotonergic receptors have been present for over 750 million years (Peroutka and Howell 1994). A primordial 5-HT receptor is believed to be the common ancestor of the catecholaminergic GPCRs and, based on sequence data, the catecholaminergic receptors supposedly evolved from the 5-HT_1 receptor (Peroutka and Howell 1994; Walker et al. 1996). Serotonin (Fig. 2) is the ligand for two different species of receptors, six GPCRs (5-$HT_{1, 2, 4-7}$, of which 5-HT_1 and 5-HT_2 are further subdivided) and one ligand-gated cation channel receptor (5-HT_3). In cypris larvae of the barnacle, *Balanus amphitrite*, DNA sequences have been found that are homologous to a human 5-HT_{1A} receptor (Kawahara et al. 1997). However, this barnacle serotonergic GPCR has not been functionally cloned nor has its pharmacology, e.g., affinity of different ligands and activation of a subsequent intracellular signal pathway, been established. Typically, mammalian 5-HT_1 receptors couple to G_i proteins that inhibit adenlyl cyclase. Furthermore, the gene coding for 5-HT_1 receptors bears no introns; the receptor protein has a relatively large third intracellular loop and a short C-terminal tail (Tierney 2001). Analysis of the nucleotide sequence of the barnacle serotonergic GPCR, putatively a 5-HT_1 receptor, indicated that the gene has no introns (Kawahara et al. 1997).

Studies on cypris larvae of *Megabalanus rosa* found that exogenously applied in 10 μmol L^{-1} serotonin had no effect on cement exocytosis from isolated cement glands (Okano et al. 1996). The exogenous application to cyprids of *B. amphitrite* suggested that this neurotransmitter is involved in modulating the settlement process (Yamamoto et al. 1999) as well as involved in the regulation of the metamorphic pathway (Yamamoto et al. 1996). Kon-ya et al. (1995) found, when using young cyprids (within one day of molting from the 6th-stage nauplius), which are not prone to settle (Clare et al. 1995), that a 24-hour exposure to exogenously applied serotonin, in the concentration range 0.3 and 10 μM, significantly induced attachment and metamorphosis in a concentration-dependent manner. However, when using old cyprids (7 days post ecdysis) serotonin had no inductive effect as compared to controls. Yamamoto et al. (1996) found that serotonin induced settlement at a concentration of 100 μM, but not in the range 0.1–10 μM. In the same study, the two dibenzazepines, imipramine and desipramine, and the dibenzcycloheptene, amitriptyline, were tested for their effect in cyprid settlement assays. These substances belong to the tricyclic antidepressant drugs and act in mammals by inhibiting the uptake of serotonin and noradrenaline by binding to carrier proteins and so facilitate transmission. These substances inhibited larval settlement in the 10 to 100-μM range while metamorphosis was promoted at lower concentrations (1–10 μM). Furthermore, the three substances interfered with the action of exogenously applied serotonin to produce larvae that were unable to attach to the substratum but which still underwent metamorphosis (so-called precocious metamorphosis) (Yamamoto et al. 1996).

The promotion of metamorphosis by imipramine, desipramine and amitriptyline at lower concentrations can indeed be due to the uptake-blocking action of these pharmacoactive compounds. In that case, the facilitation of transmission of serotonin (and/or noradrenaline) could result in metamorphosis without prior attachment and thus, indicate an important role for serotonin in the metamorphic pathway. Tricyclic antidepressant drugs are also known to interact with neurotransmitter receptors such as histamine receptors, G protein-coupled acetylcholine receptors (muscarinic ACh receptors) and 5-HT receptors (5-HT_2) (Feighner 1999). Therefore, it is possible that the settlement-inhibiting action seen in the range 10–100 μM is due to targets other than competitive binding to carrier proteins that transport serotonin to the nerve terminal where it is metabolized. In the same study, Yamamoto et al. (1996) also microinjected cyprids with serotonin and found that they settled to a greater extent than did control larvae. This result was consistent with whole animal experiments using serotonin dissolved in seawater but difficult to interpret, since control larvae did not settle.

Yamamoto et al. (1999) presented further investigations into the role of serotonin in the settlement of the barnacle *B. amphitrite*. In this study various agonists and antagonists were applied exogenously and attachment and metamorphosis were monitored. The serotonin agonists used were 5-HTQ, which interacts with 5-HT_3 receptors, serotonin creatinine sulfate, which is non-specific and α-methylserotonin, which interacts primarily with 5-HT_2 receptors. These compounds either had no effect or at certain concentrations, promoted attachment and metamorphosis as compared to artificial seawater (ASW) controls. The strongest promoting action was displayed by the 5-HT_3 agonist 5-HTQ. This substance caused a significant increase in attachment in concentrations ranging from 0.1–10 μM. Of other serotonin antagonists tested, the 5-HT_2 antagonists cyproheptadine (Fig. 3A) and LY-53 857 were the most efficient in reducing attachment and metamorphosis of cyprids (100 nM). Cinanserin (Fig. 3B), also a 5-HT_2 antagonist, impeded settlement at 1 μM. At 10 μM, the 5-HT_2 antagonists cyproheptadine and ketanserin evoked a significant increase in precocious metamorphosis, which was only observed at this concentration.

Yamamoto et al. (1999) concluded that 5-HT_2 receptors might be particularly prominent in the linkage between attachment and metamorphosis. 5-HT_2 receptors in mammals are coupled to an intracellular signal pathway, which increases cellular levels of DAG and IP_3. As yet, no attempt has been made to characterize the intracellular pathway involved in the settlement inhibition displayed by the 5-HT_2

A)

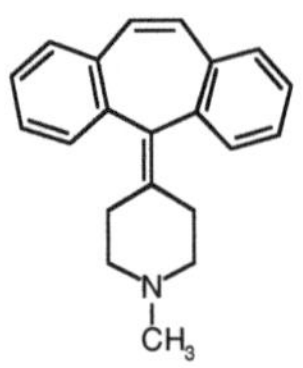

Cyproheptadine (4-(5H-Dibenzo[a,d]-cyclohepten-5-ylidine)-methylpiperidine)

5-HT_2 antagonist

EC_{50}-value (*B. amphitrite*): 100 nM

B)

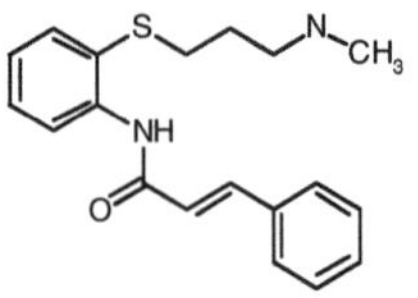

Cinanserin (N-[2-[[3-(Dimethylamino)propyl]-thio]phenyl]-3-phenyl-2-propenamide)

5-HT_2 antagonist

EC_{50}-value (*B. amphitrite*): 1 μM

Fig. 3. The chemical structures of cyproheptadine and cinanserin, their pharmacological classification in vertebrates and their EC_{50}-values in the barnacle settlement assay

antagonists. The presence of serotonin in cypris larvae was confirmed by analyzing homogenates of larvae using high performance liquid chromatography (HPLC) (Yamamoto et al. 1999). Serotonin appeared to be present at approximately 0.5 ng per cyprid.

The available evidence therefore supports the involvement of serotonin in the settlement process of the barnacle *B. amphitrite*. However, it is difficult, on the basis of existing results, to know whether serotonin is important primarily to the attachment phase or the metamorphic pathway, or indeed, both.

Directing antifouling compounds against serotonergic functions in cypris larvae could be feasible. We do believe, however, that it is desirable to gain more knowledge of the pharmacology of the serotonin receptors involved in the settlement process. Firstly, to functionally clone the receptor(s) involved and to establish its pharmacological profile, as well as the intracellular signal pathways involved.

3.2 Histamine

Histamine (2-(4-imidazolyl)-ethyl-amine) is formed by the decarboxylation of histidine. There are four known vertebrate histamine receptors i.e., H_1, H_2, H_3 and H_4, all of which are G protein-coupled receptors. Histamine mediates inflammatory responses and is present abundantly in the lungs, skin and gastrointestinal tract. It is also present in the brain where, for example, histaminergic receptors affect sedation. In invertebrates, on the other hand, the only known histaminergic receptor is a ligand-gated ion channel (Hardie 1989). This receptor is believed to be peculiar to arthropods (Roeder 2003). Evolutionarily, it is hypothesized that histamine neurotransmission appeared independently in vertebrates and invertebrates and also, that histaminergic neurotransmission appeared late in evolution (Roeder 2003). However, the overall importance of histamine for different invertebrate functions remains to be elucidated.

In barnacle cypris larvae, as well as in adult barnacles, histamine regulates the photoreceptors of the compound eye (Stuart et al. 1996, 2002). Certain levels of histamine are at all times present in the synapse but when the photoreceptor is shaded, for example by a predator, the histamine level decreases in the synapse. Histamine is thus involved in the perception of light/darkness and may accordingly be indirectly involved in the settlement process. Thus far, no vertebrate histamine agonists or antagonists have been tested in cyprid settlement assays.

3.3 γ-Aminobutyric Acid (GABA)

In mammals, GABA occurs primarily in brain tissue where it acts as the main inhibitory transmitter. GABA is formed from the decarboxylation of glutamate. There are two known types of GABA receptors: $GABA_A$, which is a ligand gated ion channel receptor and $GABA_B$, which is a G protein-coupled receptor. Studies have pointed to the importance of GABA in the settlement of a gastropod, the red abalone *Haliotis rufescens* (Morse 1990). For example, GABA mimicked the metamorphosis inducing effect of crustose coralline red algae.

The effect of GABA on barnacle settlement has been investigated by exogenous application (Yamamoto et al. 1996; Mishra and Kitamura 2000) and by microinjections (Yamamoto et al. 1996). According to Mishra and Kitamura (2000), GABA did not affect settlement as compared to ASW controls. However, Yamamoto et al. (1996) found a slight but significant inhibitory effect of GABA in concentrations ranging from 0.1–100 μM. This effect was, however, not concentration-dependent. The effects of GABA agonists and antagonists have not been studied in cypris larvae. In a separate study (Okano et al. 1996), 10 $\mu mol\ L^{-1}$ GABA did not affect cement release, by exocytosis, from isolated cement glands of *Megabalanus rosa* cyprids.

3.4 Octopamine and Tyramine

Octopamine and tyramine are referred to as "trace amines" as they are only found in trace amounts in vertebrates. Hitherto, it has been suggested that these amines and their corresponding receptors are not present in vertebrates and thus, do not play any major physiological role (Roeder 1999). However, a recent report by Borowsky et al. (2001) has demonstrated the presence of G protein-coupled receptors that are related to but distinct from the classical biogenic amine receptors such as the serotonergic and catecholaminergic GPCRs. The receptors, which were termed TA_1 and TA_2, bind the trace amines octopamine and tyramine. The functional importance of trace amines in vertebrates may, therefore, have to be re-examined.

The importance of the monoamine octopamine in invertebrates is conspicuous. Octopamine regulates a diverse range of physiological functions and interacts with targets in peripheral organs and in sensory as well as central systems (Roeder 1999, 2002; Libersat and Plueger 2004). Octopamine modulates complex behaviors including circadian rhythms, flight metabolism, feeding, aggression and dominance (Weiger 1997; Roeder 1999; Birmingham and Tauck 2003; Libersat and Pflueger 2004). Octopamine and tyramine are derived from the amino acid tyrosine. Tyrosine is converted by tyrosine decarboxylase to tyramine. Tyramine is further converted to octopamine by the enzyme tyramine-β-hydroxylase. Hence, tyramine is both a substrate for the biosynthesis of octopamine and is itself a neurotransmitter.

Three pharmacologically distinct octopamine receptors have been described in invertebrates (Roeder 1999), all of which are G protein-coupled receptors. Thus far, eight invertebrate octopamine receptors have been functionally cloned: two from the moths *Bombyx mori* and *Heliothis virescens* (von Nickisch-Rosenegk et al. 1996), two from the mollusk *Lymnaea stagnalis* (Gerhardt et al. 1997a,b), one each from the following: *D. melanogaster* (Han et al. 1998), *Aplysia californica,* and *Aplysia kurodai* (Chang et al. 2000), the honeybee *Apis mellifera* (Grohmann et al. 2003) and the cockroach *Periplaneta americana* (Bischof and Enan 2004). While the octamine receptors (OARs) from *Bombyx* and *Heliothis* have been reported to inactivate adenylyl cyclase, the OARs from *Drosophila, Lymnaea, Aplysia* and *Periplaneta* appear to activate adenylyl cyclase. The OAR from *A. mellifera* did not affect adenylyl cyclase levels but instead increased intercellular levels of Ca^{2+} (Grohman et al. 2003).

The pharmacological classification of invertebrate tyramine receptors has been hampered by the lack of proper agonists and antagonists (Roeder 2002). There is also little information about its physiological roles as a neurotransmitter. However, tyramine receptors that are negatively coupled to adenylyl cyclase have been cloned from *D. melanogaster* (Saudou et al. 1990), *Locusta migratoria* (Vanden Broeck et al. 1995), *A. mellifera* (Blenau et al. 2000) and *C. elegans* (Rex and Komuniecki 2002). Nevertheless, it is suggested that octopaminergic and tyraminergic functions in invertebrates parallel the noradrenergic/adrenergic functions in vertebrates (Roeder 1999).

Octopamine has been assayed on the cypris larvae of the barnacle *Balanus amphitrite*, by both exogenous application and microinjections (Yamamoto et al. 1996). Octopamine has also been applied exogenously to isolated cement glands from cyprids of the barnacle *Megabalanus rosa* (Okano et al. 1996). Yamamoto et al. (1996) found a significant inhibiting effect of octopamine on cyprid settlement, with complete inhibition at 100 μM. The microinjections, used in the same study, showed a similar trend. However, there are some objections to the design of this experiment (see under serotonin), since control larvae did not settle. Okano et al. (1996) found no effect of octopamine on cement exocytosis from cement glands isolated from *M. rosa*.

There are some high affinity agonists (Nathanson 1985) and a limited number of antagonists (Roeder 1990) for studying pharmacological properties of octopamine receptors. For example, phentolamine (Fig. 4), which in vertebrates is classified as an unspecific adrenergic α-antagonist, is used as an antagonist to study binding affinities to octopamine receptors (Roeder 1999). In settlement experiments of *B. amphitrite* (Yamamoto et al. 1998), phentolamine significantly inhibited attachment but also significantly promoted metamorphosis in concentrations ranging from 0.01 to 10 mM. We found, when using cypris larvae of *B. improvisus*, that phentolamine inhibited both attachment and metamorphosis in a concentration-dependent manner between 5 and 100 μM (Dahlström et al. 2000). However, in 100 μM, all larvae died within 24–48 h. In both of the above studies, phentolamine was used in settlement studies primarily for its action on α-adrenergic receptors. Since both of these studies were performed on whole animals, the molecular target for phentolamine remains to be elucidated.

Clear evidence exists for the octopaminergic influence on various physiological functions in some members of Crustacea, e.g., lobsters (Beltz 1999). Its role in cirripedes has, however, yet to be established.

3.5 The Catecholamines

The catecholamines, dopamine, noradrenaline and adrenaline (Fig. 2) are biosynthesized from the amino acid tyrosine, but their synthetic pathways are different from those of octopamine and tyramine. Tyrosine is firstly hydroxylated by tyrosine hydroxylase to L-DOPA. The decarboxylation of L-DOPA by the enzyme DOPA decarboxylase gives rise to dopamine (Fig. 2). This pathway for dopamine synthesis is identical in vertebrates and invertebrates (Hirsch and Davidson 1981). In certain vertebrate neurons dopamine is the substrate of the enzyme dopamine β-hydroxylase, whose

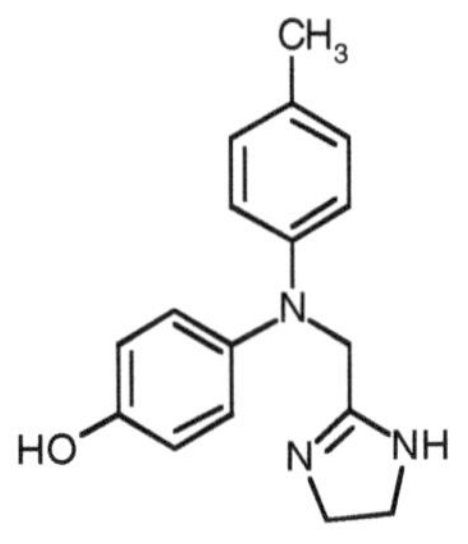

Phentolamine: 2-[N-(3-hydroxyphenyl)-p-toluidinomethyl]-2-imidazolidine

OAR antagonist, α-adrenoceptor antagonist
EC_{50}-value (*B. amphitrite*): 100 μM
EC_{50}-value (*B. improvisus*): 5 μM

Fig. 4. The chemical structure of phentolamine, pharmacological classification in vertebrates and invertebrates and its EC_{50}-values in the barnacle settlement assay

action leads to the formation of noradrenaline. Noradrenaline can be further modified by phenylethanolamine *N*-methyltransferase to yield adrenaline.

In vertebrates, dopamine modulates neuroendocrine, locomotory and emotional functions. In invertebrates, dopamine is involved in a range of physiological functions (Blenau and Baumann 2001; Liebersat and Pflueger 2004). It is found in relatively high concentrations in the insect CNS (reviewed by Blenau and Baumann 2001), as well as in the CNS of decapod crustaceans (reviewed by Beltz 1999; Tierney et al. 2003). Also, dopamine is involved in a range of behaviors such as grooming in the cockroach, *Periplaneta americana* (Weisel-Eichler et al. 1999) and the activation of courtship in the blue crab, *Callinectes sapidus* (Wood and Derby 1996). It has also been found that dopamine plays a pivotal role during development in *Drosophila melanogaster*. An increase in dopamine levels is correlated with larval molts, pupation and adult emergence (Martinez-Ramirez et al. 1992) and mutants deprived of the gene encoding the enzyme DOPA decarboxylase die early during embryogenesis (Tempel et al. 1984). Recently, it was found that dopamine levels, as measured by HPLC, fluctuated during ontogenesis in larvae of the bivalve, *Pecten maximus*, with a sharp increase at the approach of metamorphosis (Cann-Moisan et al. 2002). However, the functional significance of these changes in dopamine levels has not been established. In the barnacle, *Semibalanus balanoides*, dopamine has been shown to mimic the effect of the endogenous egg hatching signal and also, that the hatching signal caused the release of dopamine from within the embryo.

Upon release, dopamine was suggested to act as a muscle stimulant to cause egg release (reviewed by Clare 1987).

In vertebrates, dopamine receptors are classified as D1-like or D2-like and are then further subdivided. The five identified mammalian dopamine receptors are all G protein-coupled (Vallone et al. 2000). The D1-like receptors are D_1 and D_5 and activate adenylyl cyclase, giving rise to increased levels of cyclic AMP. The D2-like receptors are D_2, D_3 and D_4 and they couple to intracellular pathways that inhibit adenylyl cyclase and thus decrease intracellular levels of cyclic AMP. A number of invertebrate dopamine receptors that have been functionally cloned, e.g., from *D. melanogaster* (Gotzes et al. 1994) *A. mellifera* (Blenau et al. 1998) and *C. elegans* (Suo et al. 2002), were similar to D1-like vertebrate receptors. In addition, D2-like receptors have recently been functionally cloned from *D. melanogaster* (Hearn et al. 2002) and *C. elegans* (Suo et al. 2003).

Dopamine and the dopamine precursor L-DOPA have been used exogenously (Kon-ya and Endo 1995; Yamamoto et al. 1996, 1999) and in microinjections (Yamamoto et al. 1996) in settlement experiments on *B. amphitrite*. Furthermore, dopamine was exogenously applied to isolated cement glands of *M. rosa* (Okano et al. 1996) to study effects on cement release. L-DOPA and dopamine significantly inhibited settlement in a concentration-dependent manner at concentrations from 30 to 500 μM (Kon-ya and Endo 1995) and in the range 0.1 to 100 μM (Yamamoto et al. 1996). In addition, Kon-ya and Endo (1995) noted a marked increase in attached cyprids that failed to metamorphose in response to increased L-DOPA or dopamine concentrations. Okano et al. (1996) found that dopamine stimulated cement exocytosis and activated the directional transport of secretory granules to the sites of exocytosis. Histochemical evidence for catecholaminergic innervation was also provided by glyoxylic acid staining of isolated cement glands. Yamamoto et al. (1999) examined the effect of dopamine agonists in settlement experiments. All agonists used significantly inhibited settlement.

The D_2 agonists bromocriptine (Fig. 5 A) and *R*(-)-NPA (propylapomorphine) displayed the lowest EC_{50} values for inhibition of attachment and metamorphosis of 100 nM and 1 μM, respectively. In the same study, lisuride (Fig. 5 B), which acts as a dopamine D_2 agonist, as well as a serotonin agonist and antagonist, caused an inhibition of attachment and metamorphosis in concentrations ranging from 0.01 to 100 μM. Lisuride also significantly prolonged cyprid exploratory behaviour in concentrations ranging from 0.01–1 μM. The behavioural effect was not detected with any of the other dopamine agonists but was mimicked by mixtures of serotonin and dopamine. Moreover, using HPLC, dopamine was detected at a level of approximately 0.005 ng per cyprid.

A)

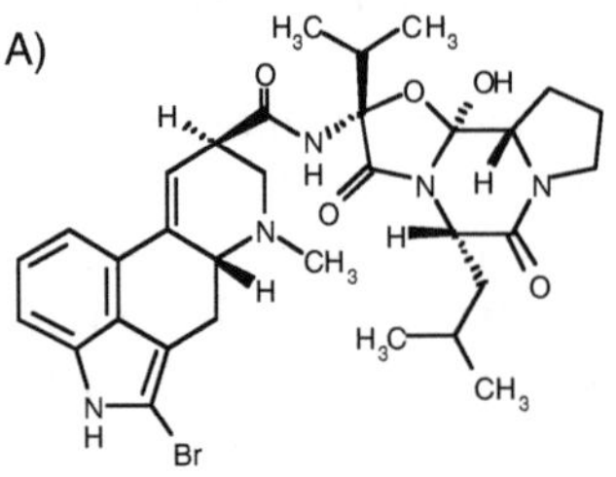

B)

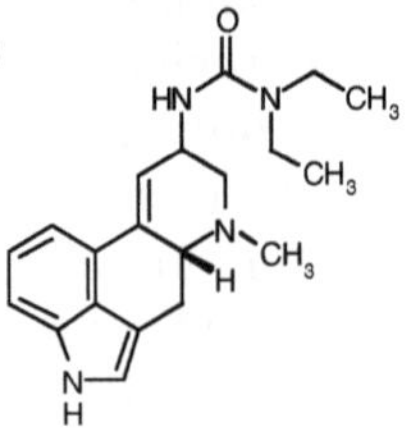

Bromocriptine ((+)-2-Bromo-12'-hydroxy-2'-(1-methylethyl)-5'-(2-methylpropyl)ergot-aman-3',6'-18-trione)

Dopamine D_2 agonist
EC_{50}-value (*B. amphitrite*): 100 nM

Lisuride (3-(9,10-didehydro-6-methyl-8-ergolinyl)-1,1-diethylurea)

Dopamine D_2 agonist, 5-HT agonist/antagonist
EC_{50}-value (*B. amphitrite*): 100 nM

Fig. 5. The chemical structures of bromocriptine and lisuride, their pharmacological classification in vertebrates and their EC_{50}-values in the barnacle settlement assay

Bromocriptine is derived from ergot alkaloids, a group of naturally occurring secondary metabolites in the fungus, *Claviceps purpurea*. Ergot alkaloids are based on the tetracyclic alkaloid lysergic acid and their pharmacological targets are quite diverse, including 5-HT receptors, adrenoceptors and dopamine receptors (Pertz 1996; Silberstein 1997). Bromocriptine is a potent D_2 agonist used for treatment in Parkinson's disease. In mammals, bromocriptine also reduces prolactin and growth-hormone secretion. Lisuride, which like bromocriptine is a synthetic ergoline, is also used for treatment in Parkinson's disease.

The role of dopamine in attachment and metamorphosis of cypris larvae has not been determined unequivocally, albeit, the results reviewed above point to the importance of dopaminergic functions in the settlement process. The receptors involved in evoking the settlement inhibition displayed by the D2-like agonists, e.g., *R*(-)-NPA and bromocriptine (Yamamoto et al. 1999), still need to be elucidated along with their pharmacological profiles and consequent intracellular pathways. However, considering the functional importance of dopamine in invertebrates, dopamine receptors might prove a promising target in barnacle larvae for novel antifouling agents.

The catecholamines, noradrenaline and adrenaline, play major roles in vertebrates and interact with adrenoceptors classified as α- and β-receptors. These are further subdivided into α_1- and α_2-receptors and β_1-, β_2- and β_3-receptors, all of which are G protein-coupled receptors. There is evidence to support the existence of at least three more subtypes each of α_1- and α_2-receptors. Noradrenaline and adrenaline have, however, hitherto only been found in trace amounts in invertebrates and their functional importance for physiological processes in invertebrates is questionable (Roeder 1999). Hence, there is to date no evidence to support noradrenergic neurotransmission before the evolutionary emergence of vertebrates. A recent report, describing catecholamine distribution in the cephalochordate, *Branciostoma lanceolatum*, did not detect noradrenaline in the nervous system (Moret et al. 2004), which supports the hypothesis that noradrenergic systems are an innovation of the vertebrate lineage.

Noradrenaline, administered exogenously, inhibited the attachment and metamorphosis of *B. amphitrite* cypris larvae at concentrations between 0.1 and 100 μM, but this result was not concentration-dependent (Yamamoto et al. 1996). Adrenaline was examined in the same study (Yamamoto et al. 1996) and did not have a significant effect on settlement. Okano et al. (1996) found that when noradrenaline was applied to isolated cement glands of *M. rosa* cypris larvae, the glands were induced to release their contents. It was also reported that noradrenaline activated the directional transport of secretory granules to the sites of exocytosis. However, the effect of noradrenaline on the cement release was weaker than that of dopamine; 10 μM of exogenously applied noradrenaline gave rise to the same effect as 1 μM of dopamine.

A *B. amphitrite* DNA sequence has been characterized with 37% homology to the human α_{2A} receptor and with particular sequence similarities to the transmembrane regions (54% homology) of this G-protein- coupled receptor (Isoai et al. 1996). However, the barnacle receptor has as yet not been functionally cloned and pharmacologically characterized and furthermore, its natural ligand remains unknown.

4 Adrenoceptor Compounds

4.1 Settlement Inhibition of *B. improvisus* Cypris Larvae

Research concerning the effect of various adrenoceptor compounds on settling *B. improvisus* cypris larvae was initiated to test the hypothesis that they were able to inhibit the release of cement used in larval permanent attachment. Okano et al. (1996) had previously shown that cement glands of *M. rosa* cyprids were induced to release their contents by exogenously applied noradrenaline or dopamine. Thus, various agonists and antagonists of G protein-coupled adrenoceptors in vertebrates were examined for their effects on the attachment and metamorphosis of *B. improvisus* cyprids.

The α_2-adrenoceptor agonists medetomidine and clonidine strongly inhibited both attachment and metamorphosis, without any significant lethal effects, in concentrations ranging from 1 nM to 80 μM (Fig. 6A–C), while the non-specific α-antagonist, phentolamine (Fig. 4), inhibited attachment and metamorphosis to a lesser degree (Dahlström et al. 2000). Phentolamine and medetomidine (Fig. 6C) were lethal to the larvae at 100 μM within 24–48 h. The EC_{50}-values of medetomidine, clonidine and phentolamine were estimated as 0.001, 0.01 and 5 μM, respectively. Medetomidine has also been examined for its ability to inhibit settlement of *B. amphitrite* cyprids and larvae of the tube-building polychaete *Hydroides elegans*. The EC_{50}-value of medetomidine on *B. amphitrite* settlement was 1 nM and identical to that for *B. improvisus*. Medetomidine displayed an EC_{50}-value of 0.1 nM for *H. elegans* settlement (Fig. 7) and an LC_{50}-value of 1 μM (Fig. 7), i.e., at 1 μM half of the larvae died within 48 h. Thus, the therapeutical range of medetomidine is similar for barnacles and *H. elegans*; however, *H. elegans* is ten times more sensitive to medetomidine.

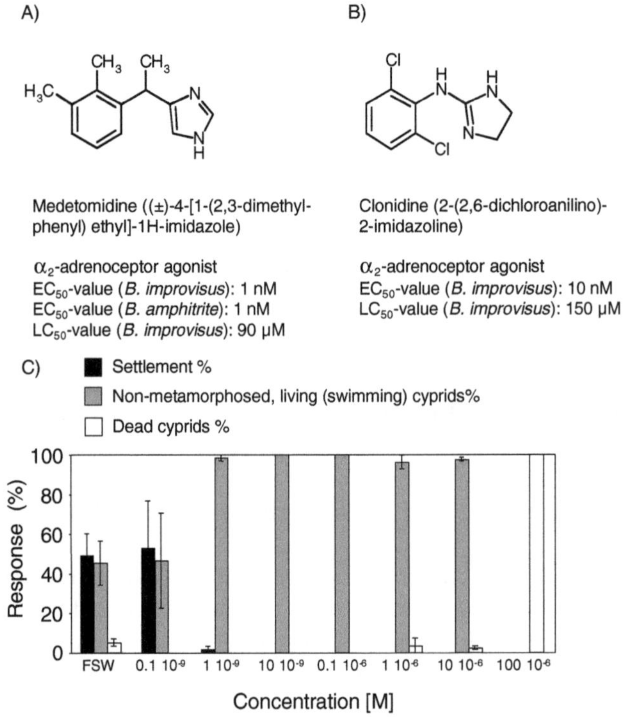

Fig. 6. The chemical structures of medetomidine (**A**) and clonidine (**B**), their pharmacological classification in vertebrates and their EC_{50}-values in barnacle settlement assays. The settlement inhibiting effect of medetomidine on *B. improvisus* cyprids (**C**). Results are presented as mean percentages ±SE (n=4). FSW, filtered seawater; *, significantly different from FSW controls. © 2000, OPA, Harwood Academic Publishers

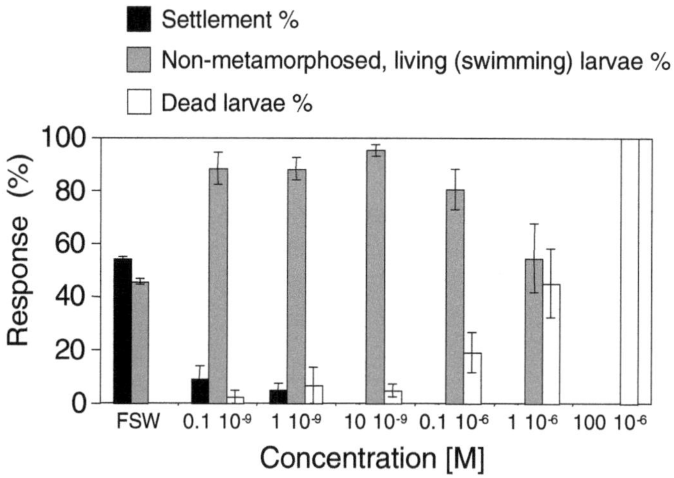

Fig. 7. The effect of medetomidine on the settlement of larvae of the tube-building polychaete *Hydroides elegans*. The EC_{50}-value of medetomidine on *H. elegans* settlement is 0.1 nM. The LC_{50}-value is 1 μM, but already at 100 nM there is a significant effect of medetomidine on the mortality of *H. elegans* larvae. Results are presented as mean percentages ±SE (n=3). FSW, filtered seawater; *, significantly different from FSW controls

In mammalian systems, medetomidine, phentolamine and clonidine interact with imidazoline binding sites (Wikberg et al. 1991; Eglen et al. 1998) as well as α-adrenoceptors. It is well established that α-adrenoceptor compounds containing imidazole, imidazoline or guanidinum groups (see Fig. 8 for chemical structures and full chemical names) are promiscuous and interact with both α-adrenergic receptors and imidazoline (I_1 and I_2) binding sites (Ernsberger 1999). Therefore, we recently conducted a further characterization of imidazoline binding site ligands, as well as α-adrenoceptor agonists, in a cyprid settlement assay (Dahlström et al. 2005). The compounds examined were chosen on the basis of their similar pharmacological classification in vertebrates and for their chemical similarities to medetomidine and clonidine (Fig. 6A, B, and Fig. 8). We found that seven of the tested compounds inhibited both attachment and metamorphosis of *B. improvisus* cyprids in a concentration-dependent and non-toxic manner from 100 nM to 10 μM (Fig. 8). Similar observations were made earlier with medetomidine, clonidine and phentolamine (Dahlström et al. 2000). However, the influence of these substances on larval searching behavior was not examined.

We further examined if these compounds interacted with the same target receptor as medetomidine. A useful tool to gain initial information on interactions at the same receptor target is to use a well-defined antagonist to abolish the effect of the agonist. Therefore, the highly selective α_2-antagonist methoxyidazoxan (Polidori et al. 2000) was used to reverse the effect of medetomidine, as well as of the compounds shown in Fig. 8. Methoxyidazoxan readily reversed the action of medetomidine when added in a 1:1 molar ratio and also, readily reversed the inhibitory action of guanabenz (α_2-agonist, I_2 ligand), moxonidine (α_2-agonist, I_1 ligand) and tetrahydrozoline (α-agonist, unspecific imidazoline binding site ligand). However, methoxyidazoxan was unable to antagonise the inhibition of settlement effected by cirazoline (I_2 ligand, α_1-agonist), metrazoline (I_2 ligand) and BU 224 (I_2 ligand).

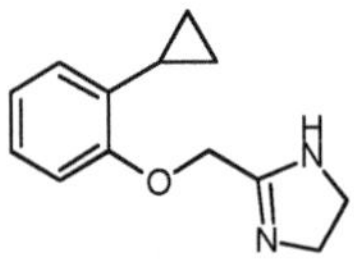

Cirazoline (2-[(2-cyclopropyl-phenoxy)methyl]-4,5-dihydro--1H-imidazole)

I_2 ligand, α_1-agonist
EC_{50}-value (*B. improvisus*): 1 μM

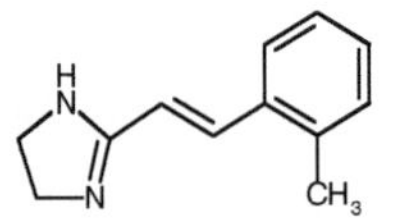

Metrazoline (4,5-dihydro-2-[(1E)-2-(2-methylphenyl)ethenyl-1H-imidazole)

I_2 ligand
EC_{50}-value (*B. improvisus*): 1 μM

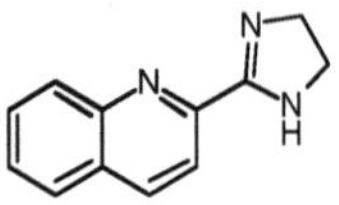

BU 224 (2-(4,5-dihydroimidazol--2-yl)quinoline)

I_2 ligand
EC_{50}-value (*B. improvisus*): 1 μM

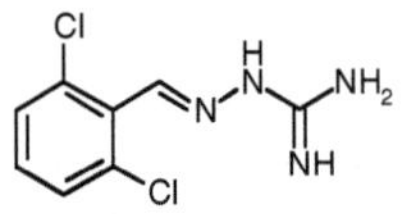

Guanabenz (1-(2,6-dichlorobenz-ylideneamino)guanidine)

α_2-agonist, I_2 ligand
EC_{50}-value (*B. improvisus*): 100 nM

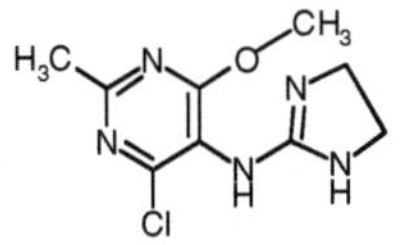

Moxonidine (4-chloro-6-methoxy-2-methyl-5-(2-imidazolin-2-yl)amino-pyrimidine)

α_2-agonist, I_1 ligand
EC_{50}-value (*B. improvisus*): 100 nM

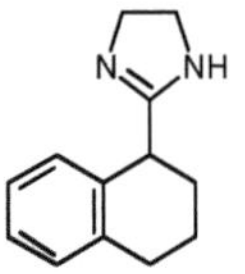

Tetrahydrozoline (1H-imidazole, 4,5-di-hydro-2-(1,2,3,4-tetrahydro-1--naphthalenyl))

Unspecific α agonist, I-site ligand
EC_{50}-value (*B. improvisus*): 1 μM

Fig. 8. The chemical structures of cirazoline, metrazoline, BU224, guanabenz, moxonidine and tetrahydrozoline, their pharmacological classification and their EC_{50}-values in the barnacle settlement assay

Thus, our results indicate that medetomidine, guanabenz, moxonidine and tetrahydrozoline interact with the same target receptor to inhibit the settlement of barnacle larvae. Furthermore, it is likely that the molecular target is a receptor with a similar pharmacological profile to adrenergic α_2-receptors. The results further suggest that the I_2 ligands cirazoline, metrazoline and BU 224 inhibit settlement of cyprids by a different mechanism to that suggested for medetomidine, moxonidine, guanabenz and tetrahydrozoline. The settlement inhibitory effect of these compounds was not reversed by methoxyidazoxan but was readily reversed upon transferring the cyprids to fresh filtered seawater (FSW), indicating that these compounds act by a non-covalent interaction with the molecular target (Dahlström et al. 2005).

In vertebrates, the I_2 binding site has not been characterized as a true receptor but is instead suggested to be an allosteric site at the enzyme monoamine oxidase (MAO) (Tesson et al. 1995), associated with mitochondrial plasma membranes. MAO exists in two forms, MAO-A and MAO-B. MAOs are the most important re-uptake mechanism for the biogenic amines noradrenaline and dopamine in vertebrates, and are thus involved in the termination of catecholamine neurotransmission. MAOs also act on serotonin and tyramine and possibly also octopamine. Therefore, I_2 ligands are now being examined as a new class of putative antidepressants for their ability to facilitate noradrenergic and serotonergic transmission (Finn et al. 2003).

The presence of MAO-A and MAO-B in invertebrates has not been fully investigated. A MAO, similar to vertebrate MAO-B, has been found in insects, but the presence and functional profile of MAO in crustaceans is poorly understood (Sloley 2004). The molecular target for BU 224, cirazoline and metrazoline in cypris larvae may be an enzyme like MAO, with implications for catecholaminergic and serotonergic neurotransmission. A classical way to investigate MAO activity is to use HPLC to identify the conversion products, i.e., acidic metabolites, of catecholamines and serotonin.

In conclusion, the studies of the different pharmacological agents and their effect on settlement of cypris larvae (Dahlström et al. 2000; 2005) imply that there are discrete molecular targets in cypris larvae, which discriminate between the structurally similar α-adrenoceptor compounds and imidazoline I_2 site ligands. The experiments have been conducted on whole animals, so information on physiological differences in the mode of action of these compounds has not been obtained. Therefore, tools for discriminating, for instance, behavioral effects on settling-stage larvae might prove useful for the pharmacological classification as well as for estimating differences in physiological effects of different receptor-active substances.

As yet, the receptor through which medetomidine exerts its action in barnacle larvae is unknown. In vertebrates, agonist binding to α_2-adrenoceptors causes an inhibition of adenylyl cyclase and thereby, decrease cytoplasmic levels of cyclic AMP. Studies are now underway to clone and functionally characterize the receptor involved in the strong settlement inhibitory action of, primarily, medetomidine, through research conducted in the Swedish Research and Development Programme Marine Paint. Also, a thorough pharmacological classification will be undertaken of this receptor, including intracellular signal pathways involved, as well as its functional expression in cypris larvae.

4.2
Surface Affinity and the Antifouling Approach

There are some common features shared by pharmaceutical formulations, i.e., systems for drug delivery, and marine antifouling paint formulations containing bioactive substances. A bioactive compound included in a marine paint has to be chemically stable and immobilized for some time within the dried paint layers. This resembles the requirements for pharmaceuticals, which need to be enclosed in different drug delivery systems, e.g., degradable polymers. When the drugs are administered, a controlled release of the active component is required (Langer 1998). This controlled release is similar to the release of bioactive components from antifouling paint formulations. A further similarity is that drugs and antifouling compounds should degrade in a controlled way. However, a major difference is the targeting of the active substance. A drug released into the body fluids usually has its target organ(s) distant from the site of release. The controlled chemistry of the surface of the delivery system is not of vital importance for drug action. In contrast, the target organisms of marine paint formulations will attempt to attach directly to the painted surface. The mechanism of how the biocide is released from the surface is thus important to the efficacy of the paint. However, any type of accumulation of the active compound at the paint surface is likely to further increase the efficacy of the paint formulation, since many biofouling organisms display a surface-bound exploratory behavior. Even if the settling propagules are passively adsorbed at the surface, a high transition of the bioactive compound at the surface will increase the antifouling effect.

A typical marine antifouling paint, containing bioactive components is based on ingredients that are less water-soluble, e.g., the binder, pigment and the biocide itself. A degradable, erosive, marine paint, based primarily on an acrylate polymer backbone, will expose its hydrophobic components to the seawater during its continuous degradation. On the other hand, polar groups, with the ability to become charged, are only present to a minor extent to control the rate of degradation. We suggest that an important feature of an effective antifouling biocide is a high degree of affinity for the interface between water and the hydrophobic paint surface (Dahlström et al. 2004). A lack of affinity might result in the rapid loss of the biocide to the water phase once the biocide has left the hydrophobic interior of the paint and reached the degradation layer, thus rendering the paint less effective. It is possible that charged interactions between the biocide and the paint surface may be important for the residence time at the surface. Here, however, there is a trade-off since a highly charged molecule is less soluble in the hydrophobic environment of the paint formulation.

4.2.1 *Surface Affinity*

The ability of an antifouling agent to adsorb to solid surfaces may have consequences for the release of the molecule from a paint coating and, as hypothesised above, for the antifouling efficacy of the paint. Optimally, the degree of affinity should guarantee a high interfacial concentration and at the same time, allow for a release of the molecule from the surface when cyprid larvae make contact. Therefore, the ability of the different adrenoceptor compounds, phentolamine, medetomidine and clonidine to adsorb to surfaces of hydrophilic and hydrophobic polystyrene (PS) was examined (Dahlström et al. 2000). The results indicated that only medetomidine could interact with both kinds of surfaces. This was evident from the total inhibition of larval settlement in dishes first incubated with medetomidine and then thoroughly rinsed before cyprids and FSW were added.

The surface adsorption of medetomidine was examined further by studying how time, pH and incubation media (milliQ water vs. FSW) affected the ability of the molecules to adsorb to the solid surfaces of both hydrophilic and hydrophobic PS (Dahlström et al. 2004). Although clonidine did not adsorb to the solid surface to the extent that inhibited settlement (Dahlström et al. 2000), it was included in the further experiments as a reference substance, since both medetomidine and clonidine inhibited cyprid settlement at similar concentrations (1–10 nM) in solution.

The results indicated that both electrostatic and hydrophobic interactions are important to the surface adsorption of medetomidine and that this compound has a stronger affinity for hydrophilic surfaces. Hydrophilic PS contains negatively charged carboxylate groups, which can form ionic interactions with the positively charged amino group of medetomidine. The pKa of medetomidine is 7.1, while that of clonidine is around 9, thus the former will be partially charged and clonidine fully charged at the normal pH of seawater (pH 8). As there is no possibility of coulombic interaction forces on the surface of the hydrophobic PS, the interaction of medetomidine with the hydrophobic dishes must have involved hydrophobic interactions. Furthermore, our results indicate that intermolecular repulsion, when the medetomidine molecules are fully charged (pH 3.6 and 5.6), does not significantly affect the surface adsorption. The high ionic strength, e.g., presence of salts in FSW, in combination with low pH (3.6) seems to alter the surface binding behavior of medetomidine to the hydrophobic surface. Settlement was not impeded in these dishes. It may be speculated that high ionic strength media, in combination with low pH, increases the self-association of the molecule, thereby reducing interface accumulation.

Since the cyprid settlement assay is a qualitative measure of the presence of the molecule at the surface, we also conducted a surface-

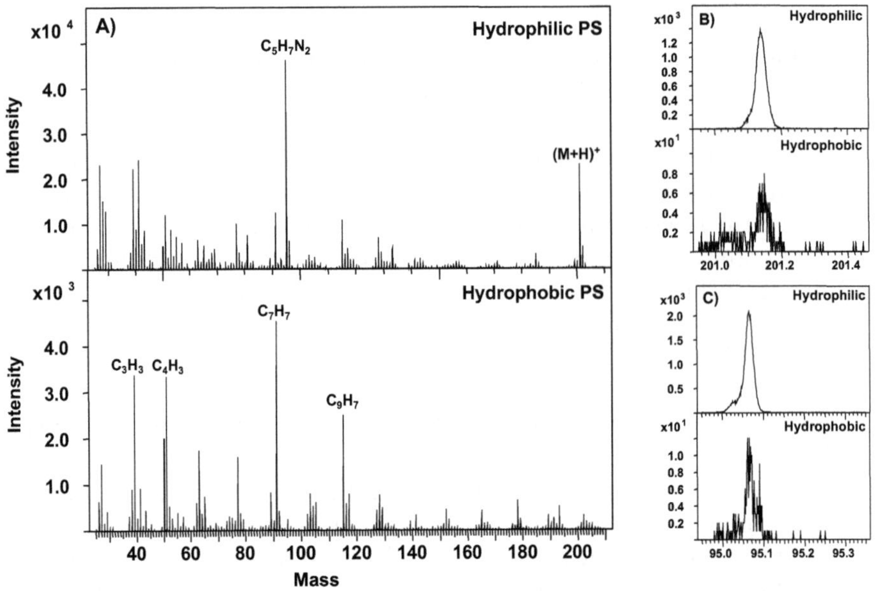

Fig. 9. Positive secondary ion mass spectra from polystyrene surfaces exposed to medetomidine. **A** survey spectrum (25 – 210 D), **B** high-resolution spectrum of the $C_5H_7N_2$ fragment ion, and **C** high-resolution spectrum from protonated medetomidine molecular ion $(M+H)^+$. © 2004 Wiley Periodicals, Inc

sensitive measurement using Time-of-Flight Secondary Ion Mass Spectrometry (TOF-SIMS). The TOF-SIMS measurements showed distinct signals of medetomidine at the hydrophilic surface, but not at the hydrophobic surface (Fig. 9). In the latter instance, medetomidine may have desorbed as TOF-SIMS is carried out under high vacuum.

Surface affinity and surface interaction are particularly important in a molecule that is targeted against the settlement of cypris larvae. The specific surface-bound settlement behavior of cyprids, which can last for several minutes without detachment (Berntsson et al. 2000; Lagersson and Hoeg 2002) make a surface-associated molecule particularly interesting. Furthermore, it has been suggested that cypris larvae are particularly adapted to perception of surface-bound chemical cues (Crisp and Meadows 1962, 1963; Gibson and Nott 1971; Clare et al. 1994). We suggest, therefore, that during exploratory behavior, cyprids encounter the medetomidine bound to the surface.

In conclusion, based on the results obtained in this study, we suggest that the high surface affinity of medetomidine is mediated through ionic and hydrophobic interactions. These interactions are strong enough to withstand rinsing, but weak enough to allow a release of the compound from the surface, thereby conveying bioavailability to exploring cyprids.

Furthermore, it is proposed that medetomidine, due to its high degree of surface affinity and considering its potential for long-term performance in the field, may serve as a lead molecule for antifouling development.

5 Conclusions

The Arthropoda is a highly diverse phylum, which contains a limited number of species that cause severe damage to human interests by destroying crops and transmitting pathogens to humans and animals. This conflict is sometimes referred to as the "arthropod pest" and is almost entirely managed by chemical insecticides, e.g., classically by *p,p'*-dichlorodiphenyl-trichloroethane (DDT) (Tedford et al. 2004). DDT, like the pyrethroid insecticides, acts on insect voltage-gated sodium channels (Söderlund and Knipple 2003). Many novel insecticides act by interaction to ligand-gated chloride channels, e.g., lindane, endosulfane and fipronil (Bloomquist 2003), e.g., the GABA chloride channel receptor. Disadvantages of these insecticides include their ability to interact with mammalian ion channel receptors and the development of tolerance in target species through modification of the GABA receptor (Ffrench-Constant et al. 1993). In several cases, it has been shown that point mutations, which replace one or two amino acids in the receptor protein, confer insecticide resistance to pest species (Miyazaki et al. 1996; Williamson et al. 1996). Therefore it has been suggested that natural toxins, e.g., the "toxin cocktails" produced by funnel-web spiders that consist of up to a hundred different peptides, could instead be explored for their potential use as natural insecticides (Tedford et al. 2004). Equivalents to the funnel web spider in the marine environment are the cone snails *Conus* sp., which are pre-eminent producers of peptidic conotoxins (Terlau and Olivera 2004). However radical the suggestion seems of using natural toxin cocktails as insecticides, it certainly puts focus on the need for working solutions for pest management in terrestrial systems.

In the marine environment, the equivalent to the 'athropod pest', namely marine biofouling caused by encrusted invertebrates, has been managed by the use of heavy metal-based coatings. The restrictions imposed on the use of these coatings have led to the increased use of so-called booster biocides, e.g., Irgarol 1051, diuron, zinc pyrithione, and Sea-Nine® 211. These compounds, like TBT, act by toxic mechanisms (Konstantinou and Albanis 2004). These biocides, like Sea-Nine® 211 [Kathon 5287 (4,5-dichloro-2-*n*-octyl-4-isothazolin-3-one)], that degrade rapidly are unlikely to persist in the marine environment (Vasishta et al. 1995; Jacobsen and Willingham 2000). Although the booster biocides have different physicochemical properties and some may degrade rapidly, there is growing concern as to the environmental fate and potential risks of their use (Thomas et al. 2002, 2003). For example, in the United Kingdom the use of Sea-Nine 211 and Irgarol 1051 has been restricted and diuron is no longer approved for use as an antifouling agent in marine paints (Thomas et al. 2002). Indeed, the need for "green chemistry" (Poliakoff et al. 2002), which addresses quality-of-life issues for man and environment, is gaining way also in marine biofouling management. Green chemistry in antifouling technologies may encompass biogenic amine receptor agonists and antagonists, which display low-toxicity, short residence time in the marine environment and display physicochemical features, which enable them to adsorb to solid surfaces, so minimizing leakage rates. Furthermore, the current knowledge on invertebrate pharmacology, e.g., the cloning of receptors and studies of their intracellular signal pathways, suggests that receptors for biogenic amines display a different pharmacology, i.e., a different profile in affinities for different synthetic agonists and antagonists, relative to their vertebrate congeneric receptors (Hen 1992). This is believed to be due to different functional constraints during the divergence of these receptors. Hence, invertebrate biogenic amine GPCRs can indeed prove good targets for biocides in antifouling technologies with low affinities for receptors of non-target species. However, more extensive investigations of, for example, barnacle receptors are necessary. Also, research regarding paint polymer composition and the interactions of pharmacoactive compounds to coating components is necessary for the design of systems, which minimize the release of biocides and, possibly, restrict their presence to the surface of the paint.

Acknowledgements. We gratefully acknowledge Professor Peter Steinberg, Dr Rocky deNys and Sophie McCloy at the Centre for Marine Biofouling and Bio-Innovation, University of New South Wales, Sydney, Australia for conducting the experiments on *H. elegans* settlement; Sonny Larsson at the Division of Pharmacognosy, Uppsala University for kind help with drawing the chemical structures of bromocriptine and lisuride, and Per Jonsson, Martin Sjögren and Kent Berntsson at Tjärnö Marine Biological Laboratory for carefully reading the manuscript. Financial support was generously provided by the Foundation for Environmental Strategic Research (MISTRA), the Foundation for Strategic Research (SSF), The Carl Trygger Foundation and the European Regional Development Fund (ERDF) through Tjärnö Centre of Excellence.

References

Alzieu CL, Sanjuan J, Deltreil JP, Borel M (1986) Tin contamination in Arcachon Bay: effects on oyster shell anomalies. Mar Pollut Bull 17:494–498

Beltz B (1999) Distribution and functional anatomy of amine-containing neurons in decapod crustaceans. Microsc Res Techn 44:105–120

Berntsson KM, Jonsson PR, Lejhall M, Gatenholm P (2000) Analysis of behavioural rejection of micro-textured surfaces and implications for recruitment by the barnacle *Balanus improvisus*. J Exp Mar Biol Ecol 251:59–83

Bicker G, Menzel R (1989) Chemical codes for the control of behaviour in arthropods. Nature 337:33–39

Birmingham JT, Tauck DL (2003) Neuromodulation in invertebrate sensory systems: from biophysics to behaviour. J Exp Biol 206:3541–3546

Bischof LJ, Enan EE (2004) Cloning, expression and functional analysis of an octopamine receptor from *Periplaneta americana*. Insect Biochem Mol Biol 34:511–521

Blenau W, Baumann A (2001) Molecular and pharmacological properties of insect biogenic amine receptors: lessons from *Drosophila melanogaster* and *Apis mellifera*. Arch Insect Biochem Physiol 48:13–38

Blenau W, Erber J, Baumann A (1998) Characterization of a dopamine D1 receptor from *Apis mellifera*: cloning, functional expression, pharmacology and mRNA localization in the brain. J Neurochem 70:15–23

Blenau W, Balfanz S, Baumann A (2000) Am*tyr*1: characterization of a gene from honeybee (*Apis mellifera*) brain encoding a functional tyramine receptor. J Neurochem 74:900–908

Bloomquist JR (2003) Chloride channels as tools for developing selective insecticides. Arch Insect Biochem Physiol 54:145–156

Borowsky B, Adham N, Jones KA, Raddatz R, Artymyshan R, Ogozalek KL, Durkin MM, Lakhlani PP, Bonini JA, Pathirana S, Boyle N, Pu X, Kouranova E, Lichtblau H, Ochoa FY, Branchek TA, Gerald C (2001) Trace amines: identification of a family of mammalian G protein-coupled receptors. Proc Natl Acad Sci USA 98:8966–8971

Brody T, Cravchik A (2000) Drosophila melanogaster G protein-coupled receptors. J Cell Biol 150:F83–F88

Cann-Moisan C, Nicolas L, Robert R (2002) Ontogenetic changes in the contents of dopamine, norepinephrine and serotonin in larvae and postlarvae of the bivalve *Pecten maximus*. Aquat Living Resour 15:313–318

Chang DJ, Li XC, Lee YS, Kim HK, Cho, N, Lo X, Weiss KR, Kandel ER, Kaang BK (2000) Activation of a heterologously expressed octopamine receptor coupled to adenylyl cyclase produces all the features of presynaptic facilitation in *Aplysia* sensory neurons. Proc Natl Acad Sci USA 97:1829–1834

Clare AS (1987) Endocrinology of cirripedes. In: Southward AJ (ed) Barnacle biology. Balkema, Rotterdam, pp 249–266

Clare AS (1996) Marine natural products antifoulants: status and potential. Biofouling 9:211–229

Clare AS, Matsumura K (2000) Nature and perception of barnacle settlement pheromones. Biofouling 15:57–71

Clare AS, Rittschof D, Gerhart DJ, Maki JS (1992) Molecular approaches to non-toxic antifouling. Invert Reprod Dev 22:67–76

Clare AS, Freet RK, McClary M (1994) On the antennular secretion of the cyprid of *Balanus amphitrite amphitrite*, and its role as a settlement pheromone. J Mar Biol Assoc UK 74:243–250

Clare AS, Thomas RF, Rittschof D (1995) Evidence for the involvement of cyclic AMP in the pheromonal modulation of barnacle settlement. J Exp Biol 198:655–664

Crisp DJ (1961) Territorial behaviour in barnacle settlement. J Exp Biol 38:429–446

Crisp DJ, Meadows PS (1962) The chemical basis of gregariousness in cirripedes. Proc R Soc Lond B 156:500–520

Crisp DJ, Meadows PS (1963) Adsorbed layers: the stimulus to settlement in barnacles. Proc R Soc Lond B 158:364–387

Dahlström M, Lindgren F, Berntsson K, Sjögren M, Mårtenssons LGE, Jonsson PR, Elwing H (2005) Evidence for different pharmacological targets for imidazoline compounds inhibiting settlement of the barnacle *Balanus improvisus*. J Exp Zool A. 303:551–562

Dahlström M, Mårtensson LGE, Jonsson PR, Arnebrant T, Elwing H (2000) Surface active adrenoceptor compounds prevent the settlement of cyprid larvae of *Balanus improvisus*. Biofouling 16:191–203

Dahlström M, Jonsson PR, Lausmaa J, Arnebrant T, Sjögren M, Holmberg K, Mårtensson LGE, Elwing H (2004) Impact of polymer surface affinity of novel antifouling agents. Biotechnol Bioeng 86:1–8

De Cesare D, Fimia GM, Sassone-Corsi P (1999) Signalling routes to CREM and CREB: plasticity in transcriptional activation. Trends Biochem Sci 24:281–285

Eglen RM, Hudson AL, Kendall DA, Nutt DJ, Morgan NG, Wilson VG, Dillon MP (1998) "Seeing through a glass darkly": casting light on imidazoline "I" sites. Trends Pharmacol Sci 19:381–390

Ernsberger P (1999) The I-1-imidazoline receptor and its cellular signaling pathways. Ann NY Acad Sci 881:35–53

Feighner JP (1999) Mechanism of action of antidepressant medications. J Clin Psychol 60:4–11

Ffrench-Constant RH, Rocheleau TA, Steichen JC, Chalmers AE (1993) A point mutation in a *Drosophila* GABA receptor confers insecticide resistance. Nature 363:449–451

Finn DP, Marti O, Harbuz MS, Vallès A, Belda X, Marquez C, Jessop DS, Lalies MD, Armario A, Nutt DJ, Hudson AL (2003) Behvioral, neuroendocrine and neurochemical effects of the imidazoline I_2 receptor selective ligand BU 224 in naïve rats and rats exposed to the stress of the forced swim test. Psychopharmacology 167:195–202

Fitzpatrick T, Ricken S, Lanzer M, Amrhein N, Macheroux P and Kappes B (2001) Subcelllular localization and characterization of chorismate synthase in the apicomplexan *Plasmodium falciparum*. Mol Microbiol 40:65–75

Gattelier L, Nagao T, Kanzaki R (2004) Serotonin modifies the sensitivity of the male silkmoth pheromone. J Exp Biol 207:2487–2496

Gerhardt C, Bakker RA, Piek GJ, Planta RJ, Vreugdenhil E, Leysen JE, van Heerikhuizen H (1997a) Molecular cloning and pharmacological characterisation of a molluscan octopamine receptor. Mol Pharmacol 51:293–300

Gerhardt CC, Lodder HC, Vincent M, Bakker RA, Plant RJ, Vreugdenhil E, Kits KS, van Heerikhuizen H (1997b) Cloning and expression of a complementary DANN encoding a molluscan octopamine receptor that couples to chloride channels in HEK 293 cells. J Biol Chem 272:6201–6207

Gibbs PE, Bryan GW, Pascoe PL, Burt GR (1990) Reproductive abnormalities in female *Ocenebra erinacea* (Gastropoda) resulting from tributyltin-induced imposex. J Mar Biol Assoc UK 70:639–656

Gibson PH, Nott JA (1971) Concerning the fourth antennular segment of the cypris larvae of Balanus balanoides. In: Crisp DJ (ed) Proc Fourth Eur Mar Biol Sym, Bangor 1969. Cambridge Univ Press, Cambridge, pp 227–236

Göpferich A (1996) Mechanisms of polymer degradation and erosion. Biomaterials 17: 103–114

Gotzes F, Balfanz S, Baumann A (1994) Primary structure and functional characterization of a *Drosophila* dopamine receptor with high homology to human $D_{1/5}$ receptors. Recept Channel 2:131–141

Grohmann L, Blenau W, Erber J Ebert PR, Strünker T, Baumann A (2003) Molecular and functional characterization of an octopamine receptor from honeybee (*Apis mellifera*) brain. J Neurochem 86:725–735

Gudermann T, Kalkbrenner F, Schultz G (1996) Diversity and selectivity of receptor-G protein interaction. Annu Rev Pharmacol Toxicol 36:429–459

Hadfield MG (2000) Why and how marine invertebrate larvae metamorphose so fast. Semin Cell Dev Biol 11:437–443

Hadfield MG, Pennington JT (1990) The nature of the metamorphic signal and its internal transduction in larvae of the nudibranch *Phestilla sibogae*. Bull Mar Sci 46:455–464

Hamm HE (1998) The many faces of G protein signalling. J Biol Chem 273:669–672

Han KA, Millar NS, Davis RL (1998) A novel octopamine receptor with preferential expression in *Drosophila* mushroom bodies. J Neurosci 18:3650–3658

Hardie RC (1989) A histamine-activated chloride channel involved in neurotransmission at a photoreceptor synapse. Nature 339:704–706

Hearn MG, Ren Y, McBride EW, Reveillaud I, Beinborn M, Kopin AS (2002) A *Drosophila* dopamine 2-like receptor: molecular characterization and identification of multiple alternatively spliced variants. Proc Natl Acad Sci USA 99:14554–14559

Hen R (1992) Of mice and flies – commonalities among 5-HT receptors. Trends Pharmacol Sci 13:160–165

Hill CA, Fox AN, Pitts RJ, Kent LB, Tan PL, Chrystal MA, Cravchik A, Collins FH, Robertson HM, Zwiebel LJ (2002) G protein-coupled receptors in *Anopheles gamibiae*. Science 298:176-178

Hirsch J, Davidson N (1981) Isolation and characterization of the dopa decarboxylase gene of *Drosophila melanogaster*. Mol Cell Biol 1:475–485

Holm ER, Nedved BT, Carpizo-Ituarte E, Hadfield MG (1998) Metamorphic-signal transduction in *Hydroides elegans* (Polychaeta: Serpulidae) is not mediated by a G protein. Biol Bull 195:21–29

International Maritime Organization (IMO) (1999) Resolution A. Antifouling systems on ships, 895 (21)

Isoai A, Kawahara H, Okazaki Y, Shizuri Y (1996) Molecular cloning of a new member of the putative G protein-coupled receptor gene from barnacle, *Balanus amphitrite*. Gene 175:95–100

Jacobson AH, Willingham GL (2000) Sea-Nine antifoulant: an environmentally acceptable alternative to organotin antifoulants. Sci Total Environ 258:103–110

Kawahara H, Isoai A, Shizuri Y (1997) Molecular cloning of a putative serotonin receptor gene from barnacle, *Balanus amphitrite*. Gene 184:245–250

Konstantinou IK, Albanis TA (2004) Worldwide occurrence and effect of antifouling paint booster biocides in the aquatic environment: a review. Environ Int 30:235–248

Kon-ya K, Endo M (1995) Catecholamines as settlement inducers of barnacle larvae. J Mar Biotechnol 2:79–81

Kon-ya K, Miki W, Endo M (1995) L-Tryptophan and related compounds induce larval settlement of the barnacle *Balanus amphitrite* Darwin. Fish Sci 61:800–803

Lacoste A, Malham SK, Gelebart F, Cueff A, Poulet SA (2002). Stress-induced immune changes in the oyster *Crassostrea gigas*. Dev Comp Immunol 26:1–9

Lagersson NC, Hoeg JT (2002) Settlement behavior and antennulary biomechanics in cypris larvae of *Balanus amphitrite* (Crustacea: Thecostraca: Cirripedia). Mar Biol 141:513–526

Langer R (1998) Drug delivery and targeting. Nature 392:5–10

Leitz T (1998) Induction of metamorphosis of the marine hydrozoan *Hydractinia echinata* Fleming 1828. Biofouling 12:173–187

Libersat F, Pflueger H-J (2004) Monoamines and the orchestration of behaviour. BioScience 54:17–25

Martinez-Ramirez AC, Ferré J, Silva FJ (1992) Catecholamines in *Drosophila melanogaster*: DOPA and dopamine accumulation during development. Insect Biochem Mol Biol 22:491–494

Meeusen T, Mertens I, De Loof A, Schoofs L (2003) G protein-coupled receptors in invertebrates: a state of the art. Int Rev Cytol 230:189–261

Mishra JK, Kitamura H (2000) The effect of mono-amino acids on larval settlement of the barnacle, *Balanus amphitrite* Darwin. Biofouling 14:299–303

Miyazaki M, Ohyama K, Dunlap DY, Matsumura F (1996) Cloning and sequencing of the *para*-type sodium channel gene from susceptible and *kdr*-resistance German cockroaches (*Blatella germanica*) and the house fly (*Musca domestica*). Mol Gen Genet 252:61–68

Moret F, Guilland J-C, Coudouel S, Rochette L, Vernier P (2004) Distribution of tyrosine hydroxylase, dopamine, and serotonin in the central nervous system of amphioxus (*Branchiostoma lanceolatum*): implications for the evolution of catecholamine systems in vertebrates. J Comp Neurol 468:135–150

Morse DE (1990) Recent progress in larval settlement and metamorphosis: closing the gaps between molecular biology and ecology. Bull Mar Sci 46:465–483

Nathanson JA (1985) Phenyliminoimidazolidines. Characterization of a class of potent agonists of octopamine-sensitive adenylate cyclase and their use in understanding the pharmacology of octopamine receptors. Mol Pharmacol 28:254–268

Nathanson JA, Hunnicutt EJ, Kantham L, Scavone C (1993) Cocaine as a naturally-occurring insecticide. Proc Natl Acad Sci USA 90:9645–9648

Newman DJ, Cragg GM, Snader KM (2003) Natural products as sources of new drugs over the period 1981–2002. J Nat Prod 66:1022–1037

Okano K, Shimizu K, Satuito C G, Fusetani N (1996) Visualization of cement exocytosis in the cypris cement gland of the barnacle *Megabalanus rosa*. J Exp Biol 199:2131–2137

Palczewski K (1997) GTP-binding-protein-coupled receptor kinases. Two mechanistic models. Eur J Biochem 248:261–269

Palczewski K, Kumasaka T, Hori T, Behnke CA, Motoshima H, Fox BA, Le Trong I, Teller DC, Okada T, Stenkamp RE, Yamamoto M, Miyano M (2000) Crystal structure of rhodopsin: a G protein-coupled receptor. Science 289:739–745

Pawlik JR (1990) Natural and artificial induction of metamorphosis of *Phragmatopoma lapidosa californica* (Polychaeta: Sabellariidae), with a critical look at the effects of bioactive compounds on marine invertebrate larvae. Bull Mar Sci 46:512–536

Peroutka SJ (1994) 5-Hydroxytryptamine receptors in vertebrates and invertebrates – why are there so many? Neurochem Int 25:533–536

Peroutka SJ, Howell TA (1994) The molecular evolution of G protein-coupled receptors – focus on 5-hydroxytryptamine receptors. Neuropharmacol 33:319–324

Pertz H (1996) Naturally occurring clavines: antagonism partial agonism at 5-HT_{2A} receptors and antagonism at alpha-1 adrenoceptors in blood vessels. Planta Med 62:387–392

Poliakoff M, Fitzpatrick JM, Farren TR, Anastas PT (2002) Green chemistry: science and politics of change. Science 297:807–810

Polidori C, Gentili F, Pigini M, Quaglia W, Panocka I, Massi M (2000) Hyperphagic effect of novel compounds with high affinity for imidazoline I_2 binding sites. Eur J Pharmacol 392:41–49

Rex E, Komuniecki RW (2002) Characterization of a tyramine receptor from *Caenorhabditis elegans*. J Neurochem 82:1352–1359

Rittschof D (1985) Oyster drills and the frontiers of chemical ecology: unsettling ideas. Am Malacol Bull 1:111–116

Rittschof D, Maki J, Mitchell R, Costlow JD (1986) Ion and neuropharmacological studies of barnacle settlement. Nether J Sea Res 20:269–275

Rittschof D (2000) Natural products antifoulants: one perspective on the challenges related to coatings development. Biofouling 15:119–127

Roberts F, Roberts CW, Johnson JJ, Kyle DE, Krell T et al (1998) Evidence for the shikimate pathway in apicomplexan parasites. Nature 393:801–805

Roeder T (1990) High-affinity antagonists of the locust neuronal octopamine receptor. Eur J Pharmacol 191:221–224

Roeder T (1999) Octopamine in invertebrates. Prog Neurobiol 59:533–561

Roeder T (2002) Biochemistry and molecular biology of receptors for biogenic amines in locusts. Microsc Res Techn 56:237–247

Roeder T (2003) Metabotropic histamine receptors – nothing for invertebrates? Eur J Pharmcol 466:85–90
Saudou F, Amlaiky N, Plassat JL, Borrelli E, Hen R (1990) Cloning and characterization of a *Drosophila* tyramine receptor. EMBO J 9:3611–3617
Silberstein SD (1997) The pharmacology of ergotamine and dihydroergotamine. Headache 37:S15–S25
Sloley BD (2004) Metabolism of monoamines in invertebrates: the relative importance of monoamine oxidase in different phyla. Neurotoxicology 25:175–183
Söderlund DM, Knipple DC (2003) The molecular biology of *knockdown resistance* to pyrethroid insecticides. Insect Biochem Mol Biol 33:563–577
Stuart AE, Morgan JR, Mekeel HE, Kempter E, Callaway JC (1996) Selective, activity-dependent uptake of histamine into an arthropod photoreceptor. J Neurosci 16: 3178–3188
Stuart AE, Mekeel HE, Kempter E (2002) Uptake of the neurotransmitter histamine into the eyes of larvae of the barnacle (*Balanus amphitrite*). Biol Bull 202:53–60
Suo S, Sasagawa N, Ishiura S (2002) Identification of a dopamine receptor from *Caenorhabditis elegans*. Neurosci Lett 319:13–16
Suo S, Sasagawa N, Ishiura S (2003) Cloning and characterization of a *Caenorhabditis elegans* D2-like dopamine receptor. J Neurochem 86:869–878
Tedford HW, Sollod BL, Maggio F, King GF (2004) Australian funnel-web spiders: master insecticide chemists. Toxicon 43:601–618
Tempel BL, Livingstone MS, Quinn WG (1984) Mutations in the dopa decarboxylase gene affect learning in *Drosophila*. Proc Natl Acad Sci USA 81:3577–3581
Terlau H, Olivera BM (2004) *Conus* venoms: a rich source of novel ion channel-targeted peptides. Physiol Rev 84:41–68
Tesson F, Limon-Boulez I, Urban P, Puype M, Vandekerckhove J, Coupry I, Pompon D, Parini A (1995) Localization of I_2-imidazoline binding sites on monoamine oxidases. J Biol Chem 270:9856–9861
Thomas KV, McHugh M, Hilton M, Waldock M (2002) Antifouling paint booster biocides in UK coastal waters: inputs, occurrence and environmental fate. Sci Total Environ 293:117–127
Thomas KV, McHugh M, Hilton M, Waldock M (2003) Increased persistence of antifouling paint biocides when associated with paint particles. Environ Poll 123: 153–161
Tierney AJ (2001) Structure and function of invertebrate 5-HT receptors: a review. Comp Biochem Phys 128A:791–804
Tierney AJ, Kim T, Abrams R (2003) Dopamine in crayfish and other crustaceans: distribution in the central nervous system and physiological functions. Microsc Res Techn 60:325–335
Vallone D, Picetti R, Borrelli E (2000) Structure and function of dopamine receptors. Neurosci Biobehav Rev 24:125–132
Vanden Broeck J, Vulsteke V, Huybrechts R, de Loof A (1995) Characterization of a cloned locust tyramine receptor cDNA by functional expression in permanently transformed *Drosophila* S2 cells. J Neurochem 64:2387–2395
Vasishtha N, Sundberg DC, Rittschof D (1995) Evaluation and release rates and control of biofouling using monolithic coatings containing isothiazolone. Biofouling 9:1–16
Von Nickisch-Rosenegk E, Krieger J, Kubick S, Laage R, Strobel J, Strotmann J, Breer H (1996) Cloning of biogenic amine receptors from moths (*Bombyx mori* and *Heliothis virescens*). Insect Biochem Mol Biol 26:817–827
Walker RJ, Brooks HL, Holden-Dye L (1996) Evolution and overview of classical transmitter molecules and their receptors. Parasitology 113:S3–S33
Weiger WA (1997) Serotonergic modulation of behaviour: a phylogenetic overview. Biol Rev 72:61–95
Weinshenker D, Garriga G, Thomas JH (1995) Genetic and pharmacological analysis of neurotransmitters controlling egg laying in *C. elegans*. J Neurosci 15:6975–6985

Weisel-Eichler A, Haspel G, Libersat F (1999) Venom of a parasitoid wasp induces grooming in the cockroach. J Exp Biol 202:957–964

Wess J (1997) G-protein-coupled receptors: molecular mechanisms involved in receptor activation and selectivity of G-protein recognition. FASEB J 11:346–354

Wikberg JES, Uhlén S, Chhajlani V (1991) Medetomidine stereoisomers delineate two closely related subtypes of idazoxan (imidazoline) I-receptors in the guinea pig. Eur J Pharmacol 193:335–340

Williamson MS, Martinez-Torres D, Hick CA, Devonshire AL (1996) Identification of mutations in the house fly para-type sodium channel gene associated with *knockdown resistance* (*kdr*) to pyrethroid insecticides. Mol Gen Genet 252:51–60

Wilson S, Bergsma DJ, Chambers JK, Muir AI, Fantom KG, Ellis C, Murdock PR, Herrity NC, Stadel JM (1998) Orphan G-protein-coupled receptors: the next generation of drug targets? Br J Pharmacol 125:1387–1392

Wood DE, Derby CD (1996) Distribution of dopamine-like immunoreactivity suggests a role for dopamine in the courtship display behaviour of the blue crab, *Callinectes sapidus*. Cell Tissue Res 285:321–330

Yamamoto H, Tachibana A, Kawaii S, Matsumura K, Fusetani N (1996) Serotonin involvement in the settlement of the barnacle *Balanus amphitrite*. J Exp Zool 275:339–345

Yamamoto H, Satuito CG, Yamazaki M, Natoyama K, Tachibana A, Fusetani N (1998) Neurotransmitter blockers as antifoulants against planktonic larvae of the barnacle *Balanus amphitrite* and the mussel *Mytilus galloprovincialis*. Biofouling 13:69–82

Yamamoto H, Shimizu K, Tachibana A, Fusetani N (1999) Roles of dopamine and serotonin in larval attachment of the barnacle *Balanus amphitrite*. J Exp Zool 284:746–758

State-of-Art Methodology of Marine Natural Products Chemistry: Structure Determination with Extremely Small Sample Amounts

M. Murata, T. Oishi, M. Yoshida

Abstract. Structure elucidation studies on natural products are reviewed emphasizing extremely small sample amounts. Previous studies on insect pheromones, periplanones, and bean-originating kairomones, glycinoeclepins, are described briefly. Recent examples are selected from marine natural products such as ciguatoxin, dolastatin-3, and aurisides. A more detailed description is given of a sperm-activating and attracting factor (SAAF), which may be the smallest sample amount used in the structure elucidation of novel non-peptidic natural products. SAAF was isolated from the eggs of the ascidian, *Ciona intestinalis*, and its structure was deduced with only approximately 4 μg (6 nmol) of sample. Based upon the proposed structure, two epimers were synthesized from chenodeoxycholic acid in 17 steps, leading to the identification of SAAF as a novel sterol sulfate.

1 Introduction

The structure elucidation of natural products has advanced greatly over the last two decades. In particular, modern nuclear magnetic resonance (NMR) technologies have shortened the time required for structure analysis. An outstanding problem in the structural studies is that specimen samples may be in short of supply. Early studies of natural products required grams of pure compounds since destructive methods were the only way to approach their chemical structures. To determine the carbon skeleton, dehydration/oxidation reactions were routinely used for comparison with authentic specimens by melting points or UV/VIS spectra. In these experiments, approximately 100 mg of samples were usually required for a single reaction. After the 1970s, when high-resolution NMR instruments became affordable for chemistry laboratories, the

M. Murata, T. Oishi
Department of Chemistry, Graduate School of Science, Osaka University, Toyonka, Osaka 560-0043, Japan

M. Yoshida
Misaki Marine Biological Station, Graduate School of Science, University of Tokyo, Miura, Kanagawa 238-0225, Japan

Progress in Molecular and Subcellular Biology
Subseries Marine Molecular Biotechnology
N. Fusetani, A.S. Clare (Eds.): Antifouling Compounds

interproton connectivity of natural products became partly available, thus facilitating the structural elucidation of the hydrogen substituted moieties. In the early 1980s, two-dimensional interproton correlation spectroscopy became feasible on a commercial instrument. This landmark innovation has greatly accelerated the structural analysis of natural biomolecules. Nevertheless, the complete determination was still a formidable task for natural product chemists. In 1986, Ad Bax invented a heteronuclear multiple bond connectivity (HMBC) method, which revealed the three-bond connection beyond a heteroatom or quaternary carbon. Pulse field gradient techniques have been applied to these inverse-detected methods such as HMBC, heteronuclear multiple quantum coherence (HMQC) and heteronuclear single quantum coherence (HSQC), which greatly improves the signal to noise ratio of the spectrum. Thanks to all these technological advances, the structures of natural products have been mostly solved by 2-D NMR spectra. In addition, mass spectrometry (MS) has become a versatile tool for the structural analysis of complicated molecules. Particularly, charge-remote-fragmentation in MS/MS experiments provided valuable structural information, which led to micro-scale structure studies of some important natural products (Adams 1990; Naoki et al. 1993). In this chapter, we review some examples of structure elucidation with extremely limited samples, and relate recent structural studies of sperm-activating and attracting factor (SAAF) from ascidians.

2
Structural Determination from Terrestrial Sources

During a long history of natural product chemistry, there have been many landmark studies in structure elucidation, a major part of which is related to terrestrial or mammalian metabolites rather than those of marine origin. These will not be discussed in detail, but some representative cases illustrate the problems associated with structural elucidation with a limited amount of sample. The first example is the periplanones A (molecular weight (MW) 232, **1**) and B (MW 248, **2**), which are sex pheromones of the American cockroach. Persoons et al. (1976) isolated 200 μg of periplanone B from many thousands of insects, and proposed several working planar structures along with a most plausible one. Adams et al. (1979) and Still (1979) confirmed the structure and further determined the absolute structure by X-ray diffraction, in combination with the MTPA/CD method and total synthesis. For periplanone A, working structure, **1'**, was proposed on the basis of extensive NMR and MS (mass spectrometry) measurements with only 20 μg (86 μmol) of specimen (Talman et al. 1978). There were two problems in this structure study: the sample used was degraded by high temperature during preparative gas chromatography, and the proposed structure was based on mass and NMR spectra. The correct structure, 1, was proposed by Hauptmann and Muhlbauer (1986) using 200 µg of sample, which was finally confirmed by total synthesis by Kuwahara and Mori (1990).

The second example is glycinoeclepin A (MW 446, **3**). This compound was isolated as an ecologically relevant hatching cue (kairomone) for soybean cyst nematode. The nematodes hatch from the eggs exactly when and where their host plants take root. In 1987, Masamune's group successfully isolated 1 mg (2.2 µmol) of an active principle from 1000 kg of kidney bean roots, which had been raised in the plantation of 10^5 m^2. The purified sample was subjected to extensive NMR measurements, including two dimensional spectra with a 500 MHz spectrometer, which was arguably the best possible instrument at that time. The NMR analysis almost reached the correct structure (**3**), which was finally determined by X-ray diffraction and total synthesis. This landmark achievement indicated that structure elucidation could be performed for novel natural products with only sub-milligram amounts of sample.

There are many excellent examples of peptidic structure identification with submicromolar quantities of sample. Well-known cases include the isolation of only a few mg of thyrotoropin releasing factor from 2,500,000 sheep brains and its structural elucidation to reveal a unique structure with a pyroglutamate moiety. Schally et al. (1971) isolated 830 µg (200 nmol) of luteinizing hormone releasing hormone (LH-RH) from the hypothalami of 165,000 pigs. The purification of an active principle was carried out by extensive chromatography and preparative electrophoresis. Both N and C termini of the peptide appeared to be blocked with pyroglutamate and amide, respectively, which made the structure determination challenging. After several trials of Edman degradation and enzymatic digestion, the structure of LH-RH was determined to be nonapeptide. Nowadays with genomic information and a dramatic reduction in sample amounts necessary for Edman degradation to sub-nanomolar quantities, physiologically important peptides, such as neuronal peptides of human origin, whose supply is usually limited, have been identified with their amino-acid sequences.

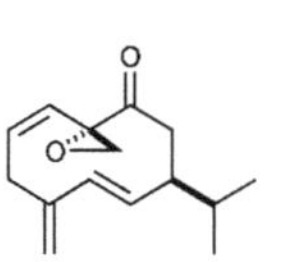

1 Periplanone-A

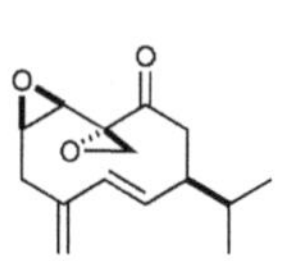

2 Periplanone-B

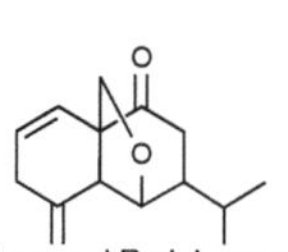

1' Proposed Periplanone-A (Incorrect)

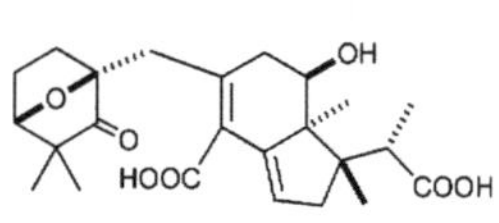

3 Glycinoeclepin-A

3 Structure Studies on Marine Natural Products

Marine natural products chemistry has a shorter history than its terrestrial counterpart. There are, however, many excellent studies of structural elucidation at micromolar scales and for complex molecules with molecular weights of over 2000 Da (Murata and Yasumoto 2000); in recent years, large molecules have often been found from marine dinoflagellates (Shimizu 1993; Yasumoto and Murata 1993). Since metabolites of marine organisms, with potent bioactivities, are often a minor constituent and the source population is occasionally limited, their structure determination is usually carried out with micromolar quantities of a specimen. This section reviews some of these studies.

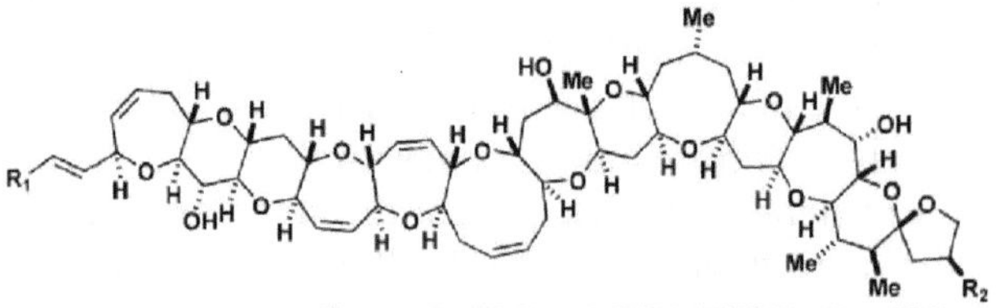

4 Ciguatoxin 1B (R_1 = $HOCH_2CH(OH)$-, R_2 = OH)
5 Ciguatoxin 4B (R_1 = CH_2=CH-, R_2 = H)

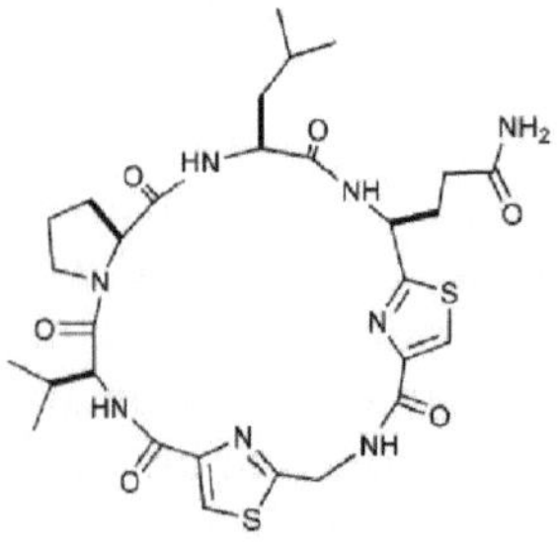

6 Dolastatin-3

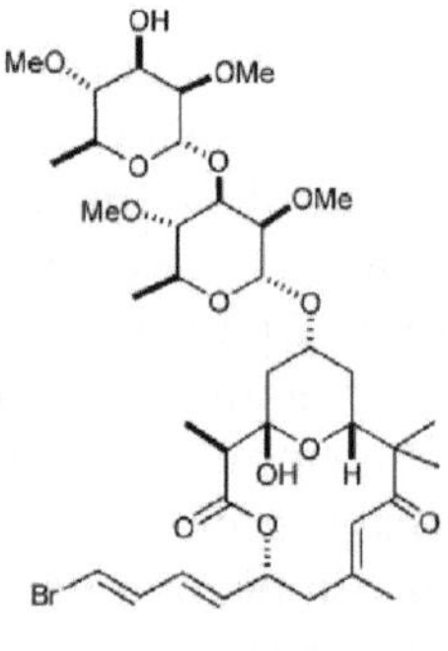

7 Auriside A

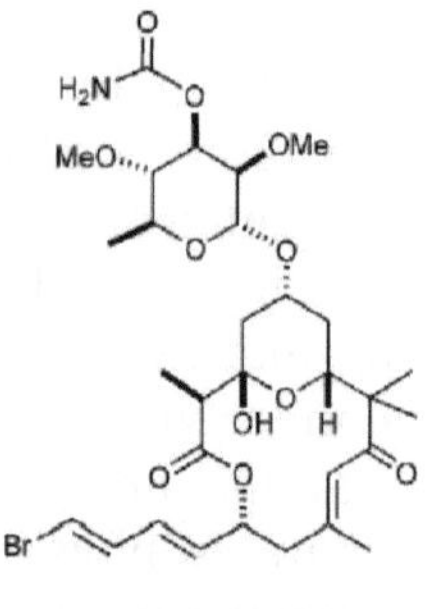

8 Auriside B

Ciguatoxin is a very potent toxin, first found in toxic coral reef fishes. The structural elucidation was carried out by Yasumoto's group in the collaboration with the Institut Louis Malardé in Tahiti (Murata et al. 1990). No more than 350 μg (0.3 μmol) of a pure specimen of ciguatoxin (CTX1B, 4, MW 1114) was subjected to extensive NMR measurements and a closely related analogue, CTX4B (5, formerly GT4b, 740 μg/0.7 μmol, MW 1064), which was isolated from a toxic dinoflagellate, was used for the confirmation of the structure. In these studies, only proton NMR data were available due to the low sensitivity of NMR instruments in the late

1980s. However, the structural features of the toxin, such as a few quaternary carbons, most of asymmetric centers residing in ring systems, and relatively large ^{1}H-^{1}H coupling constants due to a *trans*-fused ether ring, facilitated ^{1}H NMR-based structure elucidation. Other newer techniques at that time were utilized to assure the deduced structure; molecular mechanics calculations, in combination with spectral simulations, were applied to examine the interproton coupling for heavily overlapping signals, and NMR experiments at low temperatures were carried out to enhance the strength of NOE and fix the conformation of a flexible 9-membered ring. The stereochemistry at C-2 and the absolute configuration of a whole molecule was determined with the aid of synthesis. Namely, for C-2 configuration, only 5 μg of CTX1B was subjected to the reaction sequence as shown in Scheme 1, which involved protection of hydroxyl groups, ozonolysis, reduction, and derivatization for chiral resolution auxiliary (Satake et al. 1997). The absolute stereochemistry of the polycyclic ether was assigned by CD spectra in comparison between bromobenzoate of **5** and a synthetic small fragment (Fig. 1).

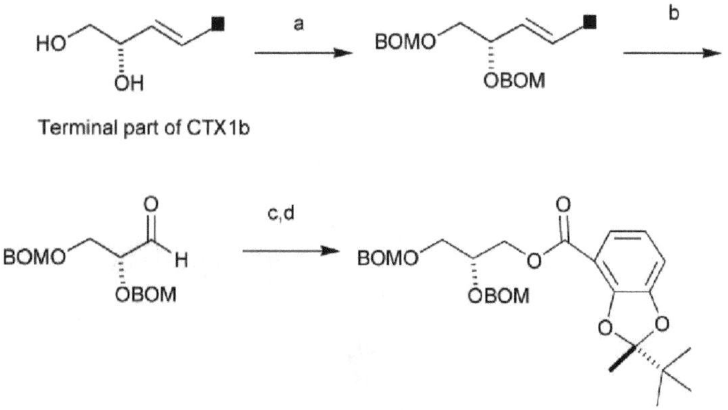

Scheme 1. Reagents and conditions: **a** BOMCI (benzyloxymethyl chloride, *i*Pr_2NEt, Bu_4NI, CH_2Cl_2); **b** OsO_4, $NaIO_4$, MeCN-H_2O (pH 7.0); **c** $NaBH_4$, MeOH; **d** (*S*)-(+)-TBMB-COCl (*S*-2-*tert*-butyl-2-methyl-1,3-benzodioxole-4-carboxyl chloride), DMAP, C_6H_6-C_5H_5N

Dolastatins are known to possess potent growth inhibitory activity against cancer cell lines. The fact that a clinical trial was attempted for the compounds indicates the potential as an anti-cancer drug, making the structure determination clinically significant. Among the dolastatins, approximately 1 mg (1.5 μmol) of dolastatin-3 (MW 660, **6**) was isolated by Pettit et al. (1982) from 100 kg of the sea hare *Dolabella auricularia*.

The sample was subjected to NMR measurements including a ^{13}C NMR spectrum and high resolution EIMS analysis. Based on these data, the structure was initially proposed to be a cyclic peptide with two thiazole rings; in particular, fragment analysis by high-resolution mass spectrometry was effectively utilized to determine the amino acid sequence. However, the initial structure, in which the valine residue was incorrectly assigned, was later revised to be **6** by NOE measurements and chiral GC analysis with a 1.8 mg dolastatin-3 obtained from newly collected sea hares. The proposed structure was further confirmed by total synthesis (Pettit et al. 1987).

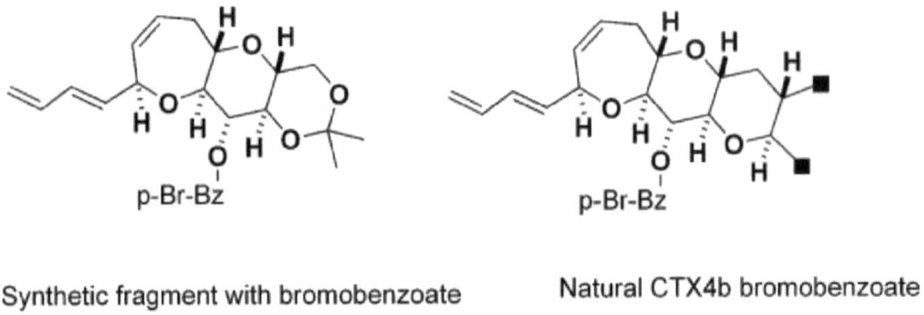

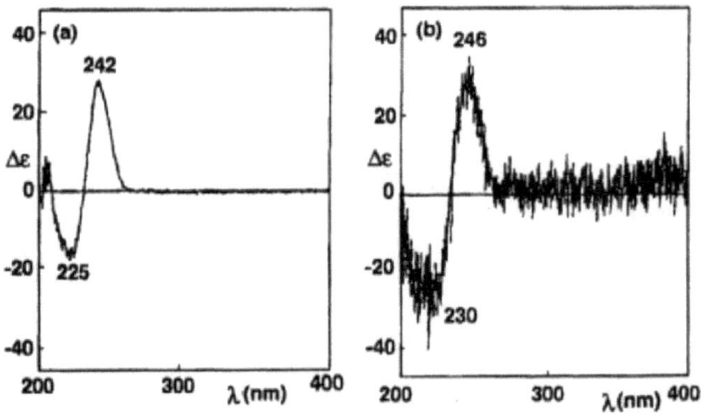

Fig. 1. CD spectra of synthetic and natural bromobenzoates. Split Cotton effects are presumably due to the interaction between the butadiene and bromobenzoate groups, thus resulting in the assignment of the absolute configuration of CTX4b (**5**) (Satake et al. 1993)

The cytotoxic agents, aurisides, were also isolated from *D. auricularia*. The structures, which were determined by Yamada's group (Sone et al. 1996), are a typical example of NMR-based analysis with only a micromolar quantity of specimen. Approximately 0.8 mg (1 μmol) of auriside A (MW 824, **7**) and 0.7 mg (1 μmol) of auriside B (MW 673, **8**) were isolated from 278 kg of sea hares. The structures were elucidated by extensive

NMR measurements such as ^{1}H-^{1}H COSY, ^{13}C-^{1}H HSQC, HMBC and NOESY. These 2D experiments are usually feasible on a modern high-field NMR instrument with a small (micromollar) sample. The relative configuration, including that between aglycon and sugar moieties, was established by NOESY and ^{1}H-^{1}H coupling constants, since the relatively rigid conformation of a macro-lactone ring facilitated the NOE-based analysis. The absolute stereochemistry was determined for a sugar moiety of rhamnose, which was hydrolyzed and derivatized for chiral HPLC comparison with synthetic specimens.

With a sample of less than 1 μmol, synthetic approaches, in addition to the spectroscopic methods mentioned above, are essential to reach the correct structure. A number of structural elucidations have been reported for marine natural products using only micromolar or submicromolar amounts of sample. Examples include cinachyrolide A (Fusetani et al. 1993), gambierol (Satake et al. 1993) and the polycavernosides (Yotsu-Yamashita et al. 1993). All of the compounds (except for aurisides) and/or their closely related congeners have been synthesized to complete their structural characterization.

4 Structure Determination of an Ascidian Sperm-Attracting and -Activating Factor (SAAF)

Quite recently, a novel sulfated sterol released from the eggs of an ascidian has been identified. This endogenous factor activates and attracts sperm upon fertilization. The structural study was carried out with an extremely small amount of no more than 6 nmol (Yoshida et al. 2002). This study is described below as an example of how to reach the correct structure with micrograms of sample.

Sperm chemotaxis, which is known for most phyla in the animal kingdom, is thought to play an essential role in fertilization (Miller 1985). The phenomenon was thus regarded as a crucial step for sperm to reach eggs, particularly in aquatic environments. The chemical factors, which may be implicated in interspecies mating and species evolution, have been the target of research for a long time. In addition, it is necessary to isolate the endogenous factors in pure form in order to investigate the mechanisms underlying sperm chemotaxis, particularly signal transduction, from chemoreception to flagellar movement. Although numerous peptides and small organic compounds were proposed as candidates for chemoattractants (Cosson et al. 1986; Punnett et al. 1992),

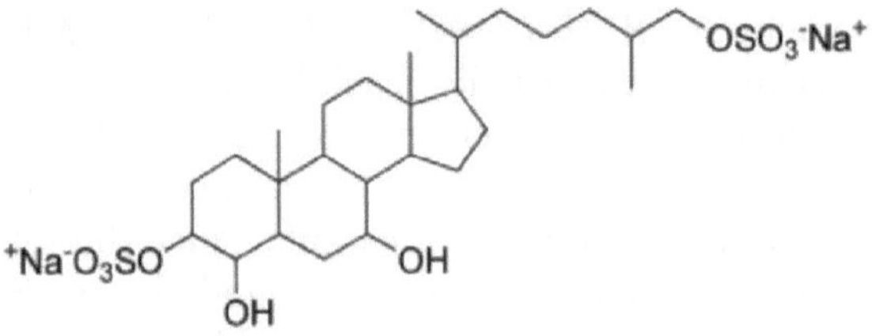

9 Proposed planer structure for SAAF

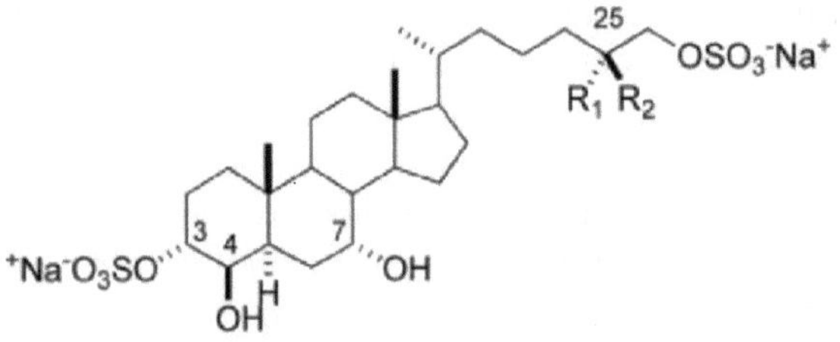

10 (25*S*): R_1=H, R_2=Me
11 (25*R*): R_1=Me, R_2=H

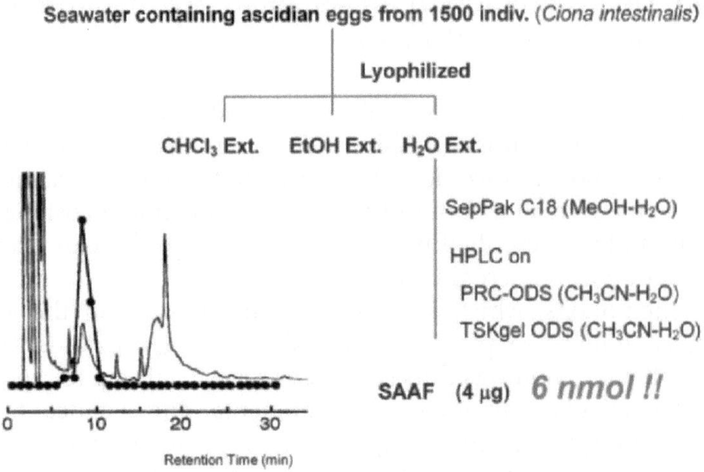

Scheme 2. Isolation of sperm-activating and -attracting factor (SAAF). The inset is the HPLC chromatograph in the final purification over a TSKgel ODS column. Dots in the chromatogram dipict the sperm-attracting activity of eluates.

only a few of them have been unambiguously identified, for example, the sea urchin *Arbacia punctulata* (Ward et al. 1985); the coral *Montipora digitata* (Coll et al. 1994) and the toad *Xenopus laevis* (Olson et al. 2001).

A difficulty in these studies is quantifying sperm chemotactic activity. Yoshida et al. (1993, 1994) carried out the pioneering study on chemotaxis in the ascidian *Ciona intestinalis*. They successfully developed a semi-quantitative method to evaluate sperm-attracting activity by monitoring sperm movements with video and by computer-aided tracking. Guided by this bioassay, they were able to extract and purify a substance which demonstrated both sperm-activating and -attracting activities as shown in Scheme **2** (Yoshida et al. 2002). Eggs were suspended in about 40 volumes of artificial seawater, and then incubated at 4°C for 14 to 20 h. After the egg suspension was centrifuged, the clear supernatant was collected and treated with ethanol and chloroform to obtain the crude SAAF fraction containing sperm-activating and -attracting activities. The crude fraction was applied onto a C_{18} reversed-phase column and eluted with a methanol-water system and then rechromatographed, under the similar conditions, with an acetonitrile–water system. The final purification was carried out by HPLC on a TSKgel reversed-phase column with a mobile phase of 21% acetonitrile. During the purification procedure, the sperm-activating and -attracting activities were always found in the same fractions, indicating that both activities were caused by a single compound. The following summarizes the structural elucidation of SAAF:

1) Isolation of SAAF (*4 μg*, 6 nmol) from seawater incubated with ascidian eggs (HPLC using H_2O-based solvent systems prevented contamination from organic solvents or containers)
2) Fast atom bombardment mass spectrum measurements (SAAF is an anionic molecule corresponding to molecular weight of 640 as Na^+ salt)
3) Electrospray time-of-flight mass spectrum measurements together with ^{1}H NMR data (Molecular formula was deduced to be $C_{27}H_{46}O_{10}S_2Na_2$)
4) ^{1}H NMR measurement in D_2O with a 1.7-mm glass tube (two up-field singlets indicated a steroid backbone of SAAF)
5) 2-D TOCSY measurements (C-3, C-4, C-6 and C-26 appeared to be oxygenated with two sulfate esters at C-3 and C-26 (see Fig. 3). The stereochemistry of the steroid skeleton was suggested to be cholestan and the configuration of oxygen substituents on rings A/B was deduced to be all axial)
6) CID MS/MS experiment (confirmation of the steroid skeleton and positions of oxygen functionalities as shown in Fig. 4)
7) Synthesis of two epimers to confirm the configuration of the working structure (one of the diastereomers with a 25*S* configuration showed an identical ^{1}H NMR spectrum to that of natural SAAF. The quantity of SAAF was determined by comparing its ^{1}H NMR spectrum with that of the synthetic product)
8) Bioassay using ascidian sperm for natural and synthetic SAAF (synthetic SAAF was as bioactive as the natural isolate)

The chemical structure of SAAF (MW 640, **10**) was chiefly deduced to be **9** using NMR and mass spectrometry. The molecular-related ions were observed in negative ion FAB MS at *m/z* 617, which corresponded to (X-Na). In addition, a prominent ion peak at *m/z* 515 corresponded to a fragment loss typical of a sulfate ester (-SO_3Na+H). These data suggested that SAAF had two sulfate ester groups; after one of them is lost, it still possessed a negatively charged group (=sulfate) showing a prominent ion at *m/z* 515. High-resolution ESI-TOF MS further supported the presence of two sulfate esters by giving rise to the exact mass of *m/z* 297.1273 (double charge), corresponding to $C_{27}H_{46}O_{10}S_2$ (calculated mass, *m/z* 297.1266). As seen above, NMR spectroscopy played a major role in deducing the structure. The ^{1}H NMR (Fig. 2) and two-dimensional ^{1}H NMR spectra (2D TOCSY) of SAAF (Fig. 3) were measured with approximately 4 μg sample; TOCSY, or other phase-sensitive spectra, appeared to give rise to better results for this amount of a sample since usual COSY experiments sometimes result in a poor S/N ratio. However, the sample was too small for measurements of the ^{13}C NMR or ^{1}H-^{13}C correlation spectrum. The 1D ^{1}H NMR spectrum revealed two singlet methyl signals at δ 0.65 and 0.97 characteristic of steroid skeletons. Two additional methyl doublets at δ 0.90 and 0.92, and complex signals at 0.8–2.0 ppm, further supported the presence of a steroid backbone. The down-field signals at δ 4.41, 3.96, 3.92, 3.83 and 3.72 corresponded to five protons on four oxygen-bearing carbon atoms (those at δ 3.92/3.83 were apparently derived from a methylene group). Among these, the connectivity of signals at δ 4.41, 3.92 and 3.72 was elucidated by 2-D TOCSY data (Fig. 3). This structural moiety fits the C-3-C-7 part of a steroid skeleton. Geminally coupled signals at δ 3.92 and 3.83 were assigned to a methylene group of C-26 on the basis of relayed correlation to one of the methyl doublets of signal. The location of sulfate groups was deduced from low-field chemical shifts of signals at δ 4.41 and 3.92/3.83 in comparison with those of hydroxyl-bearing methine and methylene groups.

High-energy collision-induced dissociation (CID) MS/MS is a versatile tool for structural elucidation of natural products with anionic functionalities (Naoki et al. 1993). The tandem MS instrument with the negative-ion fast atom bombardment (FAB) ionization method was adopted, as was the case with other sulfated compounds such as amphidinol (Satake et al. 1991), maitotoxin (Murata et al. 1994), and penarolide sulfate (Nakao et al. 2000). The product ion spectrum was recorded for precursor ions at *m/z* 515, which corresponded to a desulfated product of SAAF. The structural basis of the assignment for major product ions is given in Fig. 4. These fragmentations are reasonably

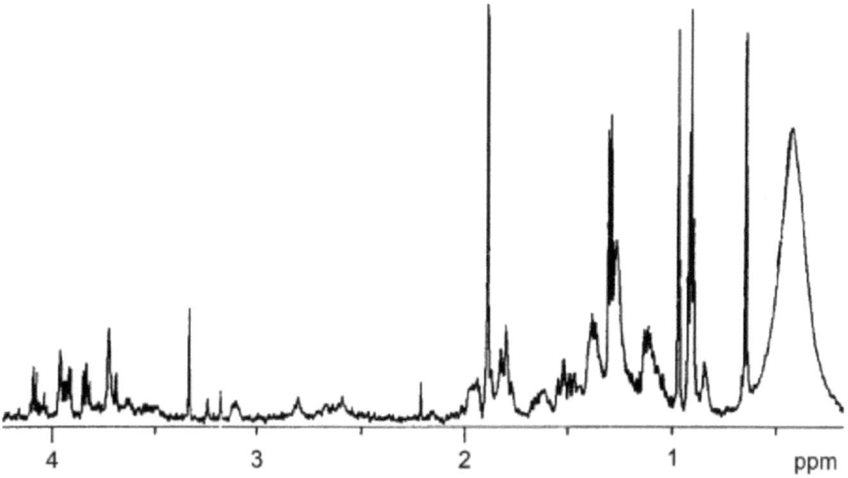

Fig. 2. ^{1}H NMR spectrum of SAAF (10, ca. 4 μg); D_2O 20 μL, 500 MHz. The spectrum was measured using a special probe (Nanolac, Z-Spec-SMIDG500) for a 1.7-mm diameter sample tube on a JEOL L-500 spectrometer. The broad peak at δ 0.4 is due to silicon grease outside of the tube used for sliding the tube into a spinner

explained by the proposed structure, particularly for the part of three oxygen functionalities. For chemical synthesis of the proposed structure, information on the configuration of SAAF is essential. The skeletal stereostructure was implied to be cholestane for the following two reasons: (1) Most of sterols isolated so far from tunicates possess a cholestane skeleton and (2) a large coupling constant between H-5 and H-6 suggested the *trans* fusion of rings A and B. The stereochemistry of oxygenated carbon atoms was deduced from the small coupling constants for the signals at δ 4.41, 3.96, and 3.72 as seen in the attached 1-D spectrum of Fig. 3. The data clearly revealed that no diaxial coupling was involved in these spin systems, which means that all the three oxygen atoms were axially oriented. The stereochemistry at C-25, which could not be assigned by NMR spectroscopy, remained to be elucidated by synthesizing both epimers at this position (**10** and **11**).

5 Synthesis of Sperm-attracting and -activating Factor and its Epimer

The synthesis of SAAF and its C-25 epimer is outlined in Scheme 3. A versatile route to synthesize both diastereomers (**10** and **11**) was realized by introducing the side chain in the latter steps. Chenodeoxycholic acid **12**

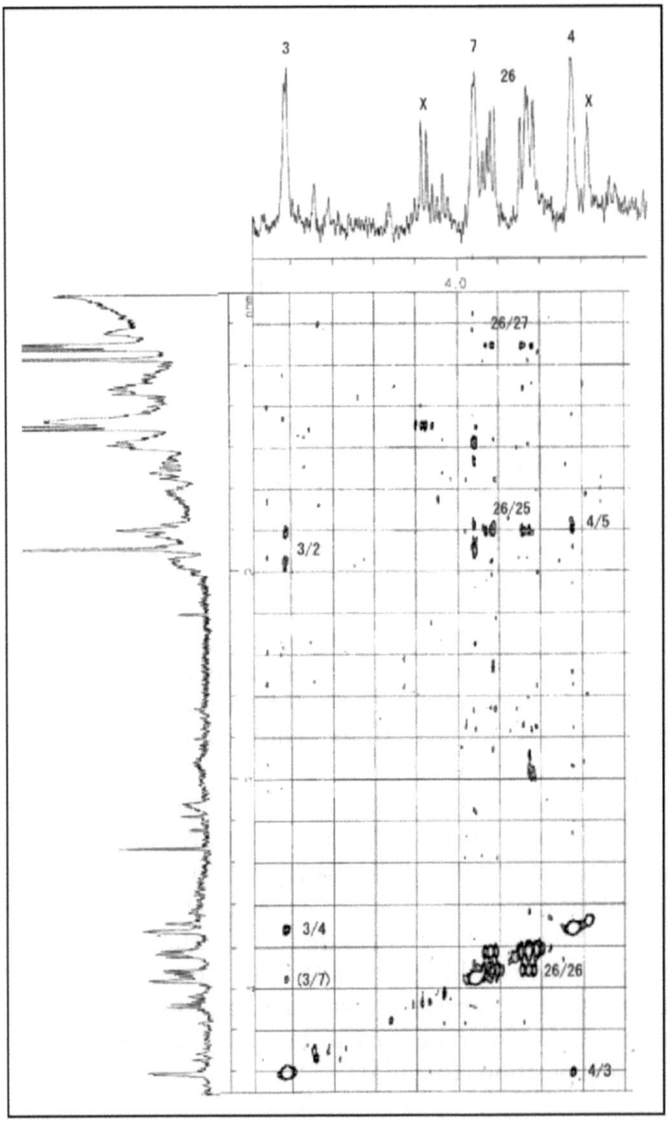

Fig. 3. TOCSY spectrum of an extremely limited amount (ca. 4 μg) of SAAF (10). The spectrum was measured using the same conditions as those in Fig. 2

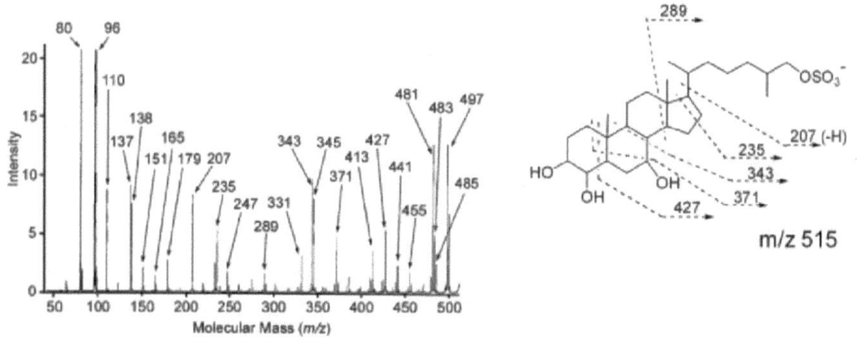

Fig. 4. CID MS/MS spectrum of SAAF (10) and fragment patterns for 3-O-defulfated ion of SAAF

was converted to known 3-keto ester **13** (Tserng 1978), which was subjected to the regioselective auto-oxidation under the basic conditions to yield 3-keto-4-enol. Selective protection of the C-7 alcohol as BOM (benzyloxymethyl) ether afforded **14**. Although a number of attempts to reduce the 3-keto-4-enol **14** using conventional methods were unsuccessful due to the formation of undesired diastereomers, a solid-phase reduction of **14** with $NaBH_3CN$ absorbed in silica gel turned out to afford desired diol **15a** in moderate yield as a mixture of C-5 epimer **15b** in a 3:1 ratio. The contiguous stereogenic centers at C-3, C-4, and C-5 on the steroid framework were then installed properly. The mixture of diols **15a** and **15b** was converted to the corresponding cyclic sulfate, and the desired **16a** was isolated from the 5β-epimer (**16b**). Regioselective opening of the cyclic sulfate at the C-3 position was achieved by treating with benzoic acid in the presence of Cs_2CO_3 in DMF (Gao and Sharpless 1988; Kim and Sharpless 1989). The methyl ester **17** was converted to aldehyde **18**; a common precursor of the two diastereomers at C-25. Elongation of the side chain of **18** was successfully achieved by a Wittig reaction using an ylide generated from (*R*)-phosphonium salt **19** to give olefin **20** ($E:Z = 1:8$) (Trost and Pulley 1995; Pettit et al. 2001). Treatment of diol **21** with SO_3·Py provided the corresponding bis-sulfate. Hydrogenation of the double bond and concomitant removal of the BOM groups afforded 25*S*-isomer (**10**). The C-25 epimer **11** was synthesized by the identical procedure to **10** except for the use of (*S*)-phosphonium salt in place of **19** (Oishi et al. 2003).

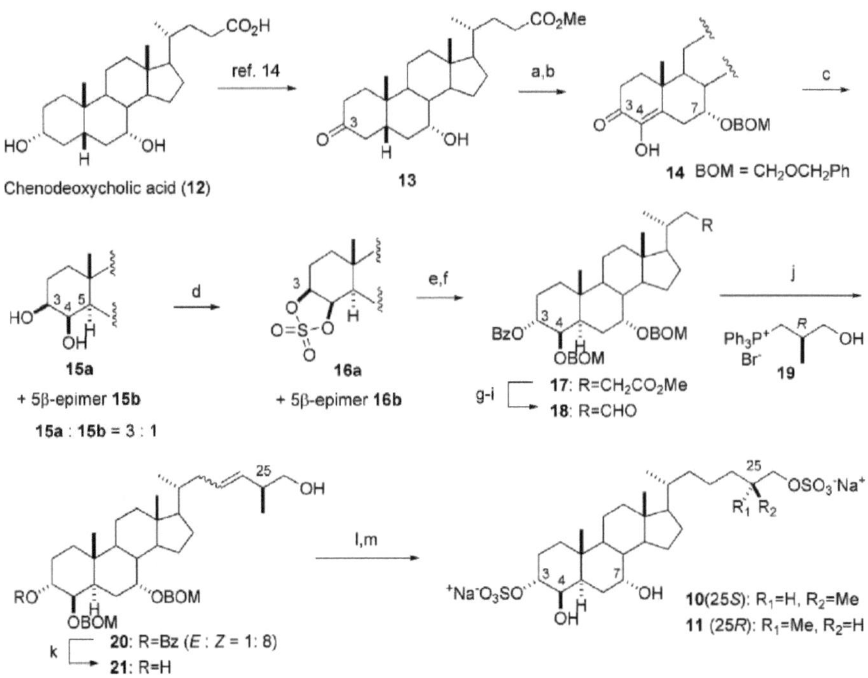

Scheme 3. Reagents and conditions: **a** *t*-BuOK, O_2, *t*-BuOH, then CH_2N_2, Et_2O, MeOH, $CHCl_3$ (71%); **b** BOMCl *i*-Pr_2NEt, CH_2Cl_2 (60%); **c** $NaBH_3CN$; $CHCl_3$, MeOH, then evaporated, 44%; **d** $SOCl_2$, Et_3N, THF, 93%, then $RuCl_3 \cdot nH_2O$, NaIO4, **16a** (65%), **16b** (11%); **e** $PhCO_2H$, Cs_2CO_3, DMF, 75%; **f** BOMCl, *i*-Pr_2NEt, CH_2Cl_2 (84%); **g** *t*-BuOK, *t*-BuOH, 95%; **h** $Pb(OAc)_4$, I_2, hv, CCl_4, 84%; **i** DMSO, 2,4,6-collidine, 94%; **j** **19**, *n*-BuLi, TMSCl, THF, then 0.5 N HCl, 78%; **k** $LiAlH_4$, THF, 80%; **l** $SO_3 \cdot Py$, Py, then Amberite IR-120B Na^+ form; **m** Pd/C, H_2, MeOH

The ^1H NMR spectrum of the natural product was compared with those of the synthetic samples, **10** (25*S*) and **11** (25*R*) (Oishi et al. 2003). While the chemical shifts of the 25-methyl group and H-26a in **11** do not match those of the natural product ($\Delta\delta$ values for H-26a and 25-methyl are –0.014 and –0.007 ppm, respectively), those of **10** are identical with SAAF, as are other resonances of the steroid framework. Thus, the configuration of the side chain was confirmed as 25*S*, resulting in the first synthesis of SAAF (Oishi et al. 2003). When the bioactivity and biosynthetic rationality are taken into account, the absolute stereochemistry is assigned as 3*R*, 4*R*, 7*R*, and 25*S*. Both the sperm-activating and -attracting activities of synthetic SAAF were bioassayed using methods previously reported (Yoshida et al. 2002). Synthetic SAAF (**10**) activated sperm of the ascidian, *Ciona intestinalis*, at 3.7 nM and concurrently exhibited the attracting activity at <10 nM; these quantitative evaluations were first accomplished with the synthetic sample (Oishi et al. 2003). The SAAF (10 nM) was delivered from an aqueous gel in a capitally tube from which SAAF diffused to a sperm-containing medium. Thus, the minimum active concentration is thought to be in the sub-nano to sub-pico molar range. It is noteworthy that 25-*epi*-SAAF (**11**) possesses comparable activity to SAAF (activated at ~3.7 nM and attracted at <10 nM) (Oishi et al. 2004).

When compared with the examples described in the former part of this section, the amount of SAAF used (6 nmol) was far lower. This may be partly due to improvement in sensitivity of modern NMR instruments, as demonstrated by the special probe for a capillary tube. Nevertheless, the high purity of the specimen seems to be more crucial. Usually, it is very difficult to prepare micrograms of an NMR sample without contamination from both HPLC and NMR solvents or even from containers. In this study, the sample was mostly treated as an aqueous solution in plastic vessels, which helped to reduce contamination by detergents and plasticizers. The only prominent contaminant seen in Fig. 2 is lactic acid; a common impurity of aqueous environments. Another problem often occurring in 2D NMR measurements is that an enormous solvent peak results in an uneven base plane. In this study, the volume of the solution was only 20 μL, which effectively reduced the contamination and secured the flatness of the base line or base plane in NMR measurements. The other methodology useful for structure analysis in the nanomolar range is MS/MS; in particular, high-energy collision experiments are helpful since C-C bond cleavages provide the information about skeletal structures. Key steps in this study were to obtain the reliable working structures and to synthesize these effectively.

6
Conclusion and Outlook

Some of the studies described in this review may be landmark achievements in the field of structure determination of natural products; particularly, the present study on SAAF would be an outstanding case in terms of sample size used for structure elucidation. With micromolar quantities, the most important information was mostly brought about by the 2-D NMR spectra. Multi-dimensional NMR technology serves as a versatile tool in structural determination of organic compounds. However, the 2-D NMR spectra become less informative when ^{1}H NMR signals overlap, due to a repetitive structure or symmetric moieties, makes the interpretation of NMR data ambiguous. In such cases, structural data derived from mass spectrometry would be very useful. As described earlier, CID MS/MS serves as a powerful and practical method to elucidate complicated organic molecules. The newest MS instruments, such as TOF-TOF and FT-ICR-MS, may potentially serve as tools for the micro-scale structural analysis of natural products. The advance in synthetic organic chemistry is another gift for natural product chemistry. Recently, a total synthesis of complex molecules was reported concomitantly with their structural determination. With these powerful tools in hand, the structures of large and complicated natural products will be solved in a much shorter period than was the case for periplanone and glycinoeclepin. Extremely limited sample amounts (e.g., nanomolar) no longer prevent structure elucidation. Thus, bioactive, but scarce compounds, whose characterization has yet to be attempted, should become targets of chemical sciences, and this will certainly widen the interdisciplinary areas of natural product chemistry.

Acknowledgements. The authors are grateful to Professors Nobuhiro Fusetani and Shigeki Matsunaga, the University of Tokyo, for valuable information, and to Prof. Takeshi Yasumoto, Japan Food Research Laboratories, for discussion.

References

Adams J (1990) Charge-remote fragmentation: analytical application and fundamental studies. Mass Spectrom Rev 9:141–186

Adams MA, Nakanishi K, Still WC, Arnorl EV, Clardy J, Persoons CJ (1979) Sex pheromone of the American cockroach: absolute configuration of periplanone-B. J Am Chem Soc 101:2494–2498

Bax A, Summers MF (1986) ^{1}H and ^{13}C assignments from sensitivity-enhanced detection of heteronuclear multiple-bond connectivity by 2D multiple quantum NMR. J Am Chem Soc 108:2093-2094

Coll JC, Bowden BF, Meehan GV, König GM, Carroll AR, Tapiolas DM, Alino PM, Heaton A, de Nys R, Leone PA, Maida M, Aceret TL, Willis RH, Babcock RC, Willis BL, Florian Z, Clayton MN, Miller RL (1994) Chemical aspects of mass spawning in corals. I. Sperm-attractant molecules in the eggs of the scleractinian coral *Montipora digitata*. Mar Biol 118:177–182

Cosson J, Carré D, Cosson MP (1986) Sperm chemotaxis in siphonophores: identification and biochemical properties of the attractant. Cell Motil Cytoskel 6:225–228

Fusetani N, Shinoda K, Matsunaga S (1993) Cinachyrolide A: a potent cytotoxic macrolide possessing two spiroketals from marine sponge *Cinachyra* sp. J Am Chem Soc 115:3977–3981

Gao Y, Sharpless KB (1988) Vicinal diol cyclic sulfates: like epoxides only more reactive. J Am Chem Soc 110:7538–7539

Hauptmann H, Muhlbauer G (1986) Identifizierung und synthese von periplanon A. Tetrahedron Lett 27:6189–6192

Kim BM, Sharpless KB (1989) Cyclic sulfates containing acid-sensitive groups and chemoselective hydrolysis of sulfate esters. Tetrahedron Lett 30:655–658

Kuwahara S, Mori K (1990) Pheromone synthesis 124. Synthesis of both the enantiomers of Hauptmann's periplanone-A and clarification of the structure of Persoons's periplanone-A. Tetrahedron 46:8083–8092

Masamune T, Anetal M, Fukuzawa A, Takasugi M, Matsue H, Kobayashi K, Ueno S, Katsui N (1987) Glycinoeclepins, natural hatching stimuli for the soybean crst nematode, *Heterodera glycines*. I. Isolation. Bull Chem Soc Jpn 60:981-999

Masamune T, Fukuzawa A, Furusaki A, Ikura M, Matsue H, Kaneko T, Abiko A, Sakamoto N, Tanimoto N, Murai A (1987) Glycinoeclepins, natural hatching stimuli for the soybean cyst nematode, *Heterodera glycines*. II. Structural elucidation. Bull Chem Soc Jpn 60:1001-1014

Miller RL (1985) Sperm chemo-orientation in the Metazoa. In: Metz B, Monroy A (eds) Biology of fertilization. Academic Press, New York, pp 275–337

Murata M, Yasumoto T (2000) The structure elucidation and biological activities of high molecular weight algal toxins: maitotoxin, prymnesins and zooxanthellatoxins. Nat Prod Rep 17:293–316

Murata M, Legrand AM, Ishibashi Y, Fukui M, Yasumoto T (1990) Structures and configurations of ciguatoxin from the moray eel *Gymnothorax javanicus* and its likely precursor from the dinoflagellate *Gambierdiscus toxicus*. J Am Chem Soc 112:4380–4386

Murata M, Naoki H, Matsunaga S, Satake M, Yasumoto T (1994) Structure and partial stereochemical assignments for maitotoxin, the most toxic and largest natural non-biopolymer. J Am Chem Soc 116:7098–7107

Nakao Y, Maki T, Matsunaga S, van Soest RWM, Fusetani N (2000) Penarolide sulfates A_1 and A_2, new a-glucosidase inhibitors from a marine sponge *Penares* sp. Tetrahedron 56:8977–8987

Naoki H, Murata M, Yasumoto T (1993) Negative ion FAB tandem mass spectrometry for structural study on polyether compounds; structural verification of yessotoxin. Rapid Commun Mass Spectrom 7:179–182

Oishi T, Tsuchikawa H, Murata M, Yoshida M, Morisawa M (2003) Synthesis of endogenous sperm-activating and attracting factor isolated from ascidian *Ciona intestinalis*. Tetrahedron Lett 44:6387–6389

Oishi T, Tsuchikawa H, Murata M, Yoshida M, Morisawa M (2004) Synthesis and identification of an endogenous sperm activating and attracting factor isolated from the ascidian *Ciona Intestinals*; an example of nanomolar-level structure elucidation of novel natural compound. Tetrahedron 44:6971–6980

Olson JH, Xiang X, Ziegert T, Kittelson A, Rawls A, Bieber AL, Chandler DE (2001) Allurin, a 21-kDa sperm chemoattractant from *Xenopus* egg jelly, is related to mammalian sperm-binding proteins. Proc Natl Acad Sci USA 98:11205–11210

Persoons CJ, Verwiel PEJ, Ritter FJ, Talman E, Nooijen PJF, Nooijin WJ (1976) Sex pheromones of the American cockroach, *Periplaneta americana*: a tentative structure of periplanone-B. Tetrahedron Lett 24:2055–2058

Pettit GR, Kamano Y, Brown P, Gust D, Inoue M, Herald CL (1982) Structure of the cyclic peptide dolastatin 3 from *Dolabella auricularia*. J Am Chem Soc 104:905–907

Pettit GR, Kamano Y, Holzapfel CW, van Zyl W J, Tuinman AA, Herald CL, Baczynsky HL, Schmidt JM (1987) The structure and synthesis of dolastatin 3. J Am Chem Soc 109:7581–7582

Pettit GR, Melody N, Herald DL (2001) Antineoplastic agents. 450. Synthesis of (+)-pancratistatin from (+)-narciclasine as relay. J Org Chem 66:2583

Punnett T, Miller RL, Yoo B-H (1992) Partial purification and some chemical properties of the sperm chemoattractant from the forcipulate starfish *Pycnopodia helianthoides* (Brandt, 1835). J Exp Zool 262:87–96

Satake M, Murata M, Yasumoto T, Fujita T, Naok H (1991) Amphidinol, a polyhydroxy-polyene antifungal agent with an unprecedented structure, from a marine dinoflagellate, *Amphidinium klebsii*. J Am Chem Soc 113:9859–9861

Satake M, Murata M, Yasumoto T (1993) Gambierol, a new toxic polyether compound isolated from the marine dinoflagellate *Gambierdiscus toxicus*. J Am Chem Soc 115:361–362

Satake M, Morohashi A, Oguri H, Oishi T, Hirama M, Harada N, Yasumoto T (1997) The absolute configuration of ciguatoxin. J Am Chem Soc 119:11325–11326

Schally AV, Arimura A, Kastin AJ, Matsuo H, Baba Y, Redding TW, Nair RM, Debeljuk, L, White WF (1971) Gonadotropin-releasing hormone: one polypeptide regulates secretion of luteinizing and follicle-stimulating hormone. Science 173:1036–1038

Shimizu Y (1993) Microalgal metabolites. Chem Rev 93:1685–1698

Sone H, Kigoshi H, Yamada K (1996) Aurisides A and B. Cytotoxic marcolide glycosides from the Japanese sea hare *Dolabella auricularia*. J Org Chem 61:8956–8960

Still WC (1979) (+/-)-Periplanone-B. Total synthesis and structure of the sex excitant pheromone of the American cockroach. J Am Chem Soc 101:2493–2494

Talman E, Verwiel PEJ, Ritter FJ, Persoons C J (1978) Sex pheromones of the American cockroach, *Periplaneta americana*. Israel J Chem 17:227–235

Tserng K-YJ (1978) A convenient synthesis of 3-keto bile acids by selective oxidation of bile acids with silver carbonate-Celite. J Lipid Res 19:501–504

Trost BM, Pulley SR (1995) Asymmetric total synthesis of (+)-pancratistatin. J Am Chem Soc 117:10143–10144

Ward GE, Brokaw CJ, Garbers DL, Vacquier VD (1985) Chemotaxis of *Arbacia punctulata* spermatozoa to resact, a peptide from the egg jelly layer. J Cell Biol 101:2324–2329

Yasumoto T, Murata M (1993) Marine toxins. Chem Rev 93:1897–1909

Yoshida M, Inaba K, Morisawa M (1993) Sperm chemotaxis during the process of fertilization in the ascidians *Ciona savignyi* and *Ciona intestinalis*. Dev Biol 157:497–506

Yoshida M, Inaba K, Ishida K, Morisawa M (1994) Calcium and cyclic AMP mediate sperm activation, but Ca^{2+} alone contributes sperm chemotaxis in the ascidian, *Ciona savignyi*. Dev Growth Differ 36:589–595

Yoshida M, Murata M, Inaba K, Morisawa M (2002) A chemoattractant for ascidian spermatozoa is a sulfated steroid. Proc Natl Acad Sci USA 99:14831–14836

Yotsu-Yamashita M, Haddock RL, Yasumto T (1993) Polycavernoside A: A novel clycosidic marolide from the red alga *Polycavernosa tsudai*. J Am Chem Soc 115:1147–1148

Subject Index

D

E

F